CAMBRIDGE LIBRARY COLLECTION

Books of enduring scholarly value

Religion

For centuries, scripture and theology were the focus of prodigious amounts of scholarship and publishing, dominated in the English-speaking world by the work of Protestant Christians. Enlightenment philosophy and science, anthropology, ethnology and the colonial experience all brought new perspectives, lively debates and heated controversies to the study of religion and its role in the world, many of which continue to this day. This series explores the editing and interpretation of religious texts, the history of religious ideas and institutions, and not least the encounter between religion and science.

Boston Monday Lectures Biology

Boston Monday Lectures Biology, a book of popular essays by the American orator Joseph Cook first published in 1879, was derived from a successful lecture series at Boston's Tremont Temple in 1878 that expertly synthesised the scientific scholarship of the day for public consumption and attempted to show that science was in harmony with religion and the Bible. Writing with clarity and conveying excitement to the lay audiences who flocked to hear him, Cook's lectures became extremely popular around the world. Biology focuses on evolution, immortality and materialism. In 13 lectures, Cook discusses topics including T.H. Huxley and John Tyndall's ideas on evolution, Rudolf Hermann Lotze's thoughts on theism, and microscopy. Cook's lectures on immortality all begin with 'Does Death End All?' before probing further into a philosophical aspect of immortality. Cook interjects short essays, which he calls 'preludes', on subjects as diverse as political patronage and Daniel Webster's death.

Cambridge University Press has long been a pioneer in the reissuing of out-of-print titles from its own backlist, producing digital reprints of books that are still sought after by scholars and students but could not be reprinted economically using traditional technology. The Cambridge Library Collection extends this activity to a wider range of books which are still of importance to researchers and professionals, either for the source material they contain, or as landmarks in the history of their academic discipline.

Drawing from the world-renowned collections in the Cambridge University Library, and guided by the advice of experts in each subject area, Cambridge University Press is using state-of-the-art scanning machines in its own Printing House to capture the content of each book selected for inclusion. The files are processed to give a consistently clear, crisp image, and the books finished to the high quality standard for which the Press is recognised around the world. The latest print-on-demand technology ensures that the books will remain available indefinitely, and that orders for single or multiple copies can quickly be supplied.

The Cambridge Library Collection will bring back to life books of enduring scholarly value (including out-of-copyright works originally issued by other publishers) across a wide range of disciplines in the humanities and social sciences and in science and technology.

Boston Monday Lectures Biology

With Preludes on Current Events

Joseph Cook

CAMBRIDGE UNIVERSITY PRESS

Cambridge New York Melbourne Madrid Cape Town Singapore São Paolo Delhi

Published in the United States of America by Cambridge University Press, New York

www.cambridge.org
Information on this title: www.cambridge.org/9781108004190

© in this compilation Cambridge University Press 2009

This edition first published 1879
This digitally printed version 2009

ISBN 978-1-108-00419-0

This book reproduces the text of the original edition. The content and language reflect the beliefs, practices and terminology of their time, and have not been updated.

Cambridge University Press wishes to make clear that the reissue of out-of-copyright books not originally published by Cambridge does not imply any knowledge or advocacy of the reissue project on the part of the original publisher.

BOSTON MONDAY LECTURES.

BIOLOGY,

WITH PRELUDES ON CURRENT EVENTS.

BY

JOSEPH COOK.

WITH A COPIOUS ANALYTICAL INDEX.

"Wie ausnahmslos universell die Ausdehnung, und zugleich wie völlig untergeordnet die Bedeutung der Sendung ist welche der Mechanismus in dem Baue der Welt zu erfullen hat."—HERMANN LOTZE.

London:
RICHARD D. DICKINSON, FARRINGDON STREET
1879.

INTRODUCTION.

THE object of the Boston Monday Lectures is to present the results of the freshest German, English, and American scholarships on the more important and difficult topics concerning the relation of Religion and Science.

CONTENTS.

LECTURES.

		PAGE
I.	HUXLEY AND TYNDALL ON EVOLUTION	1
II.	THE CONCESSIONS OF EVOLUTIONISTS	18
III.	THE CONCESSIONS OF EVOLUTIONISTS	25
IV.	THE MICROSCOPE AND MATERIALISM	35
V.	LOTZE, BEALE, AND HUXLEY ON LIVING TISSUES	45
VI.	LIFE, OR MECHANISM—WHICH?	57
VII.	DOES DEATH END ALL? INVOLUTION AND EVOLUTION	64
VIII.	DOES DEATH END ALL? THE NERVES AND THE SOUL	75
IX.	DOES DEATH END ALL? IS INSTINCT IMMORTAL?	89
X.	DOES DEATH END ALL? BAIN'S MATERIALISM	100
XI.	AUTOMATIC AND INFLUENTIAL NERVES	113
XII.	EMERSON'S VIEWS ON IMMORTALITY	126
XIII.	ULRICI ON THE SPIRITUAL BODY	137

PRELUDES.

		PAGE
I.	GIFT-ENTERPRISES IN POLITICS	45
II.	SAFE POPULAR FREEDOM	75
III.	DANIEL WEBSTER'S DEATH	89
IV.	CIVIL-SERVICE REFORM	100
V.	AUTHORITIES ON BIOLOGY	113
VI.	BOSTON AND EDINBURGH	126
VII.	THE GULF CURRENT IN HISTORY	137

BIOLOGY.

I.

HUXLEY AND TYNDALL ON EVOLUTION.[1]

"None of the processes of Nature, since the time when Nature began, have produced the slightest difference in the properties of any molecule. We are, therefore, unable to ascribe either the existence of the molecules, or the identity of their properties, to the operation of any of the causes which we call *natural*. The quality of each molecule gives it the essential character of a manufactured article, and precludes the idea of its being eternal and self-existent."—Professor CLERK MAXWELL, "Lecture delivered before the British Association at Bradford," in *Nature*, vol. viii. p. 441.

"There is a wider teleology which is not touched by the doctrine of evolution, but is actually based upon the fundamental proposition of evolution. The teleological and the mechanical views of Nature are not necessarily mutually exclusive. The teleologist can always defy the evolutionist to disprove that the primordial molecular arrangement was not *intended* to evolve the phenomena of the universe."—Professor T. H. HUXLEY in *The Academy* for October 1869, No. 1, p. 13.

IN 1868 Professor Huxley, in an elaborate paper in the "Microscopical Journal," announced his belief that the gelatinous substance found in the ooze of the beds of the deep seas is a sheet of living matter extending around the globe. The stickiness of the deep-sea mud, he maintained, is due to innumerable lumps of a transparent, jelly-like substance, each lump consisting of granules, coccoliths, and foreign bodies, embedded in a transparent, colourless, and structureless matrix. It was his serious claim that these granule-heaps, *and* the transparent gelatinous matter in which they are embedded, represent masses of protoplasm.

1. To this amazingly strategic and haughtily-trumpeted substance found at the lowest bottoms of the oceans Huxley gave the scientific name Bathybius, from two Greek words meaning *deep* and *life*, and assumed that it was in the past, and would be in the future, the progenitor of all the life on the planet. "Bathybius," was his language, "is a vast sheet of living matter enveloping the whole earth beneath the seas."

---

[1] The forty-sixth lecture in the Boston Monday Lectureship, delivered in the Meionaon, Oct. 2, 1876.

2. No less a man than David Friedrich Strauss, who, in 1872, wrote "The Old Faith and New," his last work, used Bathybius as a presumably triumphant keystone of the physiological portion of his argument against the belief in the supernatural.[1] This deep-sea ooze he made the bridge between the inorganic and the organic. "At least two miracles or revelations," wrote Jean Paul Richter, face to face with the French Revolution, "remain for you uncontested in this age, which deadens sound with unreverberating materials. They resemble an Old and a New Testament, and are these,—the birth of finite being and the birth of life within the hard wood of matter. In one inexplicable every other is involved, and one miracle annihilates a whole philosophy."[2] It is very noteworthy, that, according to Strauss's own final admission in 1872, miracle must be confessed to have occurred once at least at the introduction of life, unless some method of filling up the chasm between the dead and the living forms of matter can be found. Bathybius was to occupy this gap. "Huxley," wrote Strauss, "has discovered the Bathybius, a shining heap of jelly on the sea-bottom; Häckel, what he has called the Moneres, structureless clots of an albuminous carbon, which, although inorganic in their constitution, yet are all capable of nutrition and accretion. By these the chasm may be said to be bridged, and the transition effected from the inorganic to the organic. *As long as the contrast between inorganic and organic, lifeless and living nature, was understood as an absolute one, as long as the conception of a special vital force was retained, there was no possibility of spanning the chasm without the aid of a miracle.*"[3] As devout believers in Bathybius, educated men—Strauss affirmed in the name of what he mistook for German culture—could no longer be Christians. Bathybius had expelled miracle. Thus in 1868 and 1873 Bathybius was the watchword of the acutest anti-supernaturalistic discussions, and was adopted as a victorious weapon by Strauss, when, with his dying hand, he was using his last opportunity to equip his philosophy with armour. Men have trembled before Strauss's negation of the supernatural. Bathybius was his chief support of that denial. Huxley called his discovery *Bathybius Häckelii*. Ernst Häckel, well knowing what stupendous issues were at stake, elaborately applauded the discovery.

3. Great microscopists and physiologists, like Professor Lionel Beale and Dr. Carpenter, rejected Huxley's testimony on this matter of fact. Dr. Wallich, in 1869, in the "Monthly Microscopical Journal," presented evidence that the deep-sea ooze has nothing in it to confirm Huxley's views. The ship "Challenger," engaged now in deep-sea soundings, has accumulated evidence of the same sort; and at present Bathybius is a scientific myth and a by-word of derision. "Bathybius," says Professor Lionel Beale in his work on "Protoplasm" (London, 1874, pp. 110, 368, 371), which the "North British Review" well calls one of the most remarkable books of the age,

---

[1] The Old Faith and New, sect. 48.   [2] Levana, sect. 38.
[3] The Old Faith and New, sect. 48.

"instead of being a widely-extending sheet of living protoplasm, which grows at the expense of inorganic elements, is rather to be regarded as a complex mass of slime, with many foreign bodies and the *debris* of living organisms which have passed away. Numerous minute living forms are, however, still found upon it." At the meeting of the German Naturalists' Association at Hamburg, in September 1876, Bathybius was publicly interred. It was my fortune to converse for a while, lately, with Professor Dana of Yale College, when I put to him the question, "Does Bathybius bear the microscope?" He replied, "You know that, in a late number of 'The American Journal of Science and Arts,' Huxley has withdrawn his adhesion to his theory about Bathybius." Thus the ship "Challenger" has challenged the assertion with which Strauss challenged the world; and Huxley himself has left Bathybius to take its place with other ghosts of not blessed memory in the history of hasty speculation.

4. *Nevertheless*, in his New York definition of the doctrine of evolution, Professor Huxley speaks of a "gelatinous mass, which, so far as our present knowledge goes, is the common foundation of all life." As, by his own confession, no such gelatinous mass has ever been observed, his popular assertion that our "knowledge" goes "so far" as to establish that this gelatinous mass not only exists, but is the foundation of all life, is contradictory of his published retraction of his theory before scholars. The observed Bathybius having turned out to be a myth, its place is now occupied by an inferential Bathybius. The chasm between the inorganic and the organic was not bridged by the results of actual observation; but it must yet be bridged, even if only with a guess and a recanted theory. This substitution of the inferential for the observed is unscientific. A primary fault of Professor Huxley's latest definition of the basis of evolution is self-contradiction.

Huxley persists in his forced recantation in spite of all the alleged discoveries in the Bay of Biscay and the Adriatic. But the gelatinous mass, which, according to Huxley's New York Lectures, is the common foundation of all life, he defined. His words permit no doubt that he meant Bathybius and its associated forms of life, as Häckel does in similar language, and not protoplasm in the minute forms in which it exists in the living tissues of to-day. Huxley affirmed in New York, that, "if we traced back the animal and vegetable world, we should find, preceding what now exists, animals and plants not identical with them, but like them, only increasing their differences as we go back in time, and at the same time becoming simpler and simpler, until finally *we should arrive* at the gelatinous mass, which, so far as our present *knowledge* goes, is the common foundation of all life. The tendency of science is to justify the speculation that that also could be traced farther back, perhaps to the general nebulous condition of matter."[1]

Very plainly, by *this* gelatinous mass, at which we should "arrive"

---

[1] Tribune Pamphlet Report, p. 16.

by a process of investigation carried backward to the first living organisms and to the nebulous condition of matter, Huxley does *not* mean protoplasm in minute forms in the veins of the nettle, and in the other living tissues of to-day, and in them constituting what his famous lecture of a few years ago called " the physical basis of life." But he affirmed that our "knowledge," and not merely our theory, goes " so far " as to show that *this* gelatinous mass is " the foundation of all life."

In view of his recantation as to this sheet of living matter beneath the seas, this assertion is self-contradictory. Since no such gelatinous mass has ever been seen, the substitution of an inferential for an observed sheet of living slime enveloping the world is unscientific. With the argument of Huxley, that of Strauss takes its place among exploded and ludicrous errors.

5. It follows, also, from the facts now stated, that *Professor Huxley's New York Lectures are defective in omitting the most essential part of their subject; that is, in failing to explain how evolution bridges the chasm between the inorganic and the organic, or the lifeless and the living forms of matter.*

6. There have been and are at least three schools of evolutionists, —those who deny the Divine existence, those who ignore it, and those who affirm it ; or the atheistic, the agnostic, and the theistic. Carl Vogt, Büchner, and Moleschott belong to the atheistic school of evolutionists ; Huxley and Tyndall and Spencer, to the agnostic ; Dana, Gray, Owen, Dawson, Carpenter, Sir J. Herschell, Sir W. Thomson, and, in the judgment of Professor Gray, Darwin himself, to the theistic.

7. Of the theistic form of the doctrine of evolution, there are theoretically three varieties : (1) That which limits the supernatural action in the origination of species to the creation of a few primordial cells; (2) That which makes Divine action in the origination of species chiefly indirect, or through the agency of natural causes, and yet sometimes direct, or through special creation ; (3) That which makes God immanent in all natural law, and regards every result of cosmic forces as the outcome of present Divine action.

8. In the history of the discussion of evolution. the origin of species among plants and animals has been explained by at least seven distinct hypotheses :—

(1.) Self-elevation by appetency, or use and effort. That is the theory of Monboddo, Lamarck, and Cope.

(2.) Modification by the surrounding condition of the medium. That is Geoffrey St. Hillaire, Quatrefages, Draper, and Spencer.

(3.) Natural selection under the struggle for existence, with spontaneous variability, causing the survival of the fittest. That is Darwin and Häckel.

(4.) Derivation by pre-ordained succession of organic forms under an innate tendency or internal force. That is Owen and Mivart.

(5.) Evolution by unconscious intelligence. That is Morell, Laycock, and Murphy.

(6.) Immanent action and direction of Divine power, working by the purposive collocation and adjustment of natural forces, acting without breaks; or the theory of *creative evolution.* That is Asa Gray, Baden Powell, and the Duke of Argyll.

(7.) The same immanent Divine power collocating and adjusting natural forces, but with breaks of special intervention, and this notably in the case of man. That is Dana, and Darwin's great co-discoverer of evolution, Alfred Wallace.[1]

What Huxley calls the Miltonic theory of creation, he did well not to call the biblical; for it is generally admitted by specialists in exegetical science, that the writings of Moses neither fix the date, nor definitely describe the mode, of creation. Professor Dana, in the closing chapter of his celebrated "Geology," exhibits the first chapter of Genesis as thoroughly harmonious with geology, and as both true and divine. Many theologians combine their distinctive positions with some theistic view of evolution, especially with that held by Professor Dana. Owenism seems at least as sure of a future as unmodified Darwinism. Dana and Häckel represent respectively, I should say, the use and the abuse of the theory of evolution.

9. It is thus evident, from the history of recent speculation alone, that there are, or well may be, at least thirty different views as to the past history of nature; but Professor Huxley affirms that, so far as he knows, "there have been, and well can be, only three." That nature has existed from eternity, and that it arose, according to the Miltonic hypothesis, in six natural days, and that it originated by evolution, of which latter he gives a definition,—these are his three theories; and they are a curiously incomplete statement of facts in the case. It does not follow, that, if the first two be overthrown, only the theory represented by *his definition* is left to be chosen; but this is the implicit and explicit assumption of the New York Lectures.

10. It is the theistic, and not the agnostic or the atheistic, school of evolution which is increasing in influence among the higher authorities of science.

Some agnostics are proud of exhibiting under almost atheistic phraseology a really theistic philosophical tendency. Spencer's negations in natural theology amount to the assertion that our knowledge of the Divine existence is like our knowledge of the back-side of the moon,—we know that it is, not what it is. But I assuredly know that there is not a ripple on any sedgy shore, or in the open sea of the whole gleaming watery zone, from here to Japan, which is not influenced by that unknown side as much as by the known. So, in the far-flashing spiritual zones of the universe of worlds, there is not a ripple which does not owe glad allegiance to that law of moral gravitation which proceeds from the whole Divine nature, known and unknown. God is knowable, but unfathomable. The agnostics call God unknowable; but that He is unfathomable is all that they prove, and often all that they mean.

---

[1] See arts. on "Evolution," by Professor Youmans and President Seelye, in Johnson's Cyclopædia, and Johnson's Natural History.

11. As Professor Huxley does not notice the different schools of evolutionists, his New York definition of the doctrine is defective through vagueness.

12. For the same reason, it is defective by a suppressed statement of hypotheses which are rivals of his own.

13. It is evident, from the nature of the case, that the question of chief interest to religious science is, whether the new philosophy is to be established in its atheistic, its agnostic, or its theistic form. But Professor Huxley regards the order of the appearance of species as a matter to be studied with all zeal: the causes of their appearance, he thinks, are a matter of subordinate importance. At Buffalo he said, "All that now remains to be asked is, How development was effected? and that is a subordinate question." He thus makes the merely initial question, What? more important than the commanding and final question, Why? The clashing looms in Machinery Hall at the World's Exhibition are of supreme moment; the Corliss Engine, which drives them, is of subordinate and inferior interest. Religious science, therefore, finds Professor Huxley curiously wanting in the sense of logical proportion.

14. The New York Lectures insist on resemblances, and not on differences, in related animal forms.

15. They exaggerate resemblances by broadly inaccurate pictorial representation. The Eocene horse of Wyoming, of the genus *Orohippus*, Dana says, is not larger than a fox.[1] The bones of its leg and foot were represented in the New York reported illustrations as quite as large as those of the horse.

16. The New York Lectures prove the existence, not of connected links, but of links with many gaps between them. They prove the existence of steps with many and long and unexplained breaks, and should prove the existence of an inclined plane.

17. They fail to reply to the great, and as yet unanswered objections to Darwinism,—the absence of discovered links between man and the highest apes, the sterility of hybrids, the mental and moral superiority of man, and the existence, in many animals, of organs of no use to the possessors under the laws of either natural or sexual selection.

18. In asserting that this self-contradictory, vague, and historically inexact account of evolution is a demonstration of the truth of his definition, and places evolution, thus defined, on "exactly as secure a foundation" as the Copernican theory, which is verified by all experiment, and has in its favour the unanimity of experts, Professor Huxley's conclusions include more than his premises.

The New York Lectures disagree in their conclusions with those of higher geological authorities, equally well or better acquainted with the American facts, and notably with the conclusions of Dana and Verrill. According to these professors of the university where the relics are preserved, the bones explain, in part, the variations of one style, but do not account for gaps between groups of animals, and least of all do they account for man.[2]

---

[1] Manual of Geology, ed. of 1875, p. 505. [2] Dana, Manual of Geology, pp. 590–604.

Professor Gray calls himself, in his latest work, a "convinced theist, and religiously an accepter of the creed commonly called the Nicene."[1] Is there yet any occasion for the disquietude of a free mind holding these views? If the demonstrative evidence in favour of the materialistic form of the theory of evolution is unsatisfactory as presented by Huxley in New York, what shall be said of the subtler procedures of Tyndall's Belfast Address?

Sitting on the Matterhorn on a July day in 1868, Tyndall meditates on the period when the granite was a part of the molten world; thinks then of the nebula from which the molten world originated; and asks next whether the primordial formless fog contained potentially the sadness with which he regarded the Matterhorn.[2] In 1874 he answers, Yes, and concludes that we must recast our definitions of matter and force, since life and thought are the flower of both.

Accordingly, Tyndall's effort is to change the definition of matter. Of the many forms of materialism, his coincides nearest with a tendency which has been gathering strength among physicists for the last hundred years,—to deny that there are two substances in the universe, matter and mind, with opposite qualities, and to affirm that there is but one substance, matter, itself possessed of two sets of properties, or of a physical side and a spiritual side, making up a double-faced unity.[3] This is precisely the materialism of Professor Bain of Aberdeen, and of Professor Huxley; and its numerous supporters in England, Scotland, and Germany, are fond of proclaiming that among metaphysicians, as well as among physiologists, it is the growing opinion; and that the arguments to prove the existence of two substances have now entirely lost their validity, and are no longer compatible with ascertained science and clear thinking.

Tyndall's speculations as to matter are simply an extension of the hypothesis of evolution, according to the scientific doctrine of uniformity, from the known to the unknown. Back to a primordial germ Darwin is supposed by Tyndall to have traced all organisation: back to the properties of unorganised matter in a primordial nebula Tyndall now traces that germ. Evolution explains everything since the germ. Evolution must be applied to explain as much as possible before the germ. So far as we can test her processes by observation and experiment, Nature is known to proceed by the method of evolution: where we cannot test her processes, analogy requires that we should suppose that she proceeds by the same method. As all the organisations now or in past time on the earth were potentially in the primordial germ, so that germ was potentially in the unorganised particles of the primordial star-dust: in other words, there was latent in matter from the first the power to evolve organisation, thought, emotion, and will. Where matter obtained this power, or whether matter is self-existent, physical science has no means of determining. In the evolution of the universe from a primor-

---

[1] Darwiniana, 1876, p. vi.
[2] Musings on the Matterhorn, 27th July 1868. Note at end of Tyndall's Address on Scientific Materialism, 19th August 1868.
[3] Bain (Professor Alexander), Mind and Body, 1873, pp. 130, 140, 191, 196.

dial haze of matter possessing both physical and spiritual properties, there has been no design other than that implied in the original constitution of the molecular particles. Of course, it is utterly futile to oppose these views as self-contradictory in the light of the established definition of matter.

Many of the replies made to Professor Tyndall, however, miss the central point in his scheme of thought, and endeavour to show that it is madness to imagine that matter, as now and for centuries defined by science, can evolve organisation and life. But no one has proclaimed the insanity of such a supposition more vigorously than Tyndall has himself. "These evolution notions," he exclaims, "are absurd, monstrous, and fit only for the intellectual gibbet, in relation to the ideas concerning matter which were drilled into us when young."[1] Most assuredly Professor Tyndall does not propose "to sweep up music with a broom," or "to produce a poem by the explosion of a type foundry." Audacities of that sort are to be left to the La Mettries and Cabanis and Holbachs: they are not attempted even by the Büchners and Carl Vogts and Moleschotts and Du Bois Reymonds, who, with some whom Tyndall too much resembles, are now obsolete or obsolescent in Germany. "If a man is a materialist," said Professor Tholuck to me once, as we walked up and down a celebrated long arbour in his garden at Halle, "we Germans think he is not educated." In the history of speculation, so many forms of the materialistic theory have perished, that a chance of life for a new form can be found in nothing less fundamental than a change in the definition of matter. Tyndall perceives, as every one must who has any eye for the signs of the times in modern research, that if Waterloos are to be fought between opposing schools of science or between science and theology or philosophy, the majestic line of shock and onset must be this one definition. "Either let us open our doors freely to the conception of creative acts," he says in the sentence which best indicates his point of view in his Belfast Address, " or, abandoning them, let us radically change our notions of matter."

Now, it is singular, and yet not singular, that one can find nowhere in Tyndall's writings the changed definition on which everything turns. The following four propositions, all stated in his own language, taken from different parts of his recent discussions, are the best approach to a definition that I have been able to find in examining all he has ever published on materialism:—

1. "Emotion, intellect, will, and all their phenomena, were once latent in a fiery cloud."[2] "I discern in matter the promise and potency of every form and quality of life."[3] "Who will set limits to the possible play of molecules in a cooling planet? Matter is essentially mystical and transcendental."[4]

2. "Supposing that, in youth, we had been impregnated with the notion of the poet Goethe, instead of the notion of the poet Young, looking at matter not as brute matter, but as the living garment of God, is it not probable that our

---

[1] Address on the Scientific Use of the Imagination, 1870.
[2] Tyndall, Fragments of Science, Eng. ed., p. 163.
[3] Belfast Address, 1874.
[4] Tyndall, Fragments of Science, Eng. ed., p. 163.

repugnance to the idea of primeval union between spirit and matter might be considerably abated?"[1]

3. "Granting the nebula and its potential life, the question, Whence come they? would still remain to baffle and bewilder us. The hypothesis does nothing more than transport the conception of life's origin to an indefinitely distant past."[2]

4. "Philosophical defenders of the doctrine of uniformity . . . have as little fellowship with the atheist, who says that there is no God, as with the theist, who professes to know the mind of God. 'Two things,' said Immanuel Kant, 'fill me with awe: the starry heavens, and the sense of moral responsibility in man.' . . . The scientific investigator finds himself overshadowed by the same awe."[3] "I have noticed during years of self-observation that it is not in hours of clearness and vigour that the doctrine (of materialistic atheism) commends itself to my mind, and that, in the presence of stronger and healthier thought, it ever dissolves and disappears, as offering no solution of the mystery in which we dwell, and of which we form a part."[4]

Of the definition of matter implied in these extracts, it must be affirmed,—not that it is new,—for it is simply what the schools call hylozoism, modified by the recent forms of the atomic theory and of the doctrine of evolution, but that *it reverses the best established position of science.*

1. It denies, and the established definition affirms, that inertia, in the strict sense of the word, is a property of matter.
2. It affirms, and the established definition denies, that matter has power to evolve organisation and vitality.
3. It affirms, and the established definition denies, that matter has power to evolve thought, emotion, conscience, and will.

In the conflict between the established definition of matter and Tyndall's definition, I, for one, prefer the established, for the following reasons:

1. If inertia is a property of matter, the power to evolve organisation, life, and thought, cannot be; but that inertia is a property of matter is a proposition susceptible of overwhelming proof from the necessary beliefs of the mind, from common consent, from the agreement of philosophers in all ages, and from all the results of experiment and observation.

Of course, the logical existence of the alternatives implied in this argument is denied by those who attribute both inertia and spiritual properties to matter as a mystic, transcendental, double-faced unity; but, while they used the word "inertia," their definition of it is not the established one, as is that here employed. By force, I mean that which is expended in producing or resisting motion. By inertia, I mean the incapacity to originate force or motion, or that quality which causes matter, if set in motion without other resistance than itself can supply, to keep on moving for ever; or, if left at rest without other force than its own, to remain at rest for ever. Materialism, hylozoism, and Tyndall's definition of matter, cannot justify themselves, unless it be proved that inertia is not a property of matter. Every student of this theme knows, and in this presence it is

---

[1] Fragments of Science, p. 165.   [2] Ibid., p. 166.   [3] Ibid., p. 167.
[4] Additions to the Belfast Address, in Tyndall's authorised edition.

unnecessary for me to state, what the proofs are that matter cannot move itself. They are far more superabundant and crucial than even those which support the belief in the existence of gravitation. Newton himself did not regard attraction as an essential property of matter ; and it was long a debate whether his great generalisation should be named the theory of attraction, or the theory of propulsion. If the established definition of matter, and the consequent proof of the spiritual origin of all force, or of the Divine immanence in natural law, are not to be disestablished until that late day when the proof that inertia is not a property of matter, that is, that matter can move itself, can be put into the form of a syllogism, then the yoke of Socrates, Aristotle, and Plato,—of which Tyndall complains, that, after twenty centuries, it is yet unbroken,—is likely to continue to be what it now is, one of the best examples in history of the survival of the fittest.

2. The established definition of matter rests on facts verifiable by experience ; Tyndall's, confessedly, is demanded and supported only by the tendencies of an improved theory of evolution.

"Those who hold the doctrine of evolution," says Tyndall himself, "are by no means ignorant of the uncertainty of their data, and they yield no more to it than a provisional assent. They regard the nebular hypothesis as probable ; and, in the utter absence of any evidence to prove the act illegal, they extend the method of nature from the present into the past, and accept as probable the unbroken sequence of development from the nebula to the present time."[1]

In his Belfast Adress, Tyndall says, " The strength of the doctrine of evolution consists not in an experimental demonstration, but in its general harmony with the method of Nature as hitherto known." But *his definition of matter rests only on this theory*, which, as he admits, is not verified by experiment ; while the accepted definition of matter is so verified. It is notoriously to experiment, and to ages of experiment, and to necessary belief itself, that the accepted definition appeals ; it is to the exigencies of an unverified, and experimentally unverifiable theory, that Tyndall appeals.

3. According to the doctrines of analogy and uniformity, on which Tyndall relies, matter must be supposed to be inert where we cannot experiment on it, since it is where we can.

4. Tyndall admits that the manner of the connection between matter and mind is unthinkable, and that, "if we try to comprehend that connection, we sail in a vacuum." His own definition, therefore, involves propositions which are unthinkable. They must have been reached by sailing through a vacuum, and can be proved only by a similarly adventurous voyage.

Pertinent exceedingly to the criticism of his definition of matter are Tyndall's famous admissions that "molecular groupings and molecular motions explain nothing ; " that "the passage from the physics of the brain to the corresponding facts of consciousness is unthinkable ; " and that, "if love were known to be associated with a right-

---

[1] Fragments of Science, p. 166.

handed spiral motion of the molecules of the brain, and hate with a left-handed, we should remain as ignorant as before as to the cause of the motion."[1] If the connection between matter and thought in the brain is so obscure, that neither Tyndall, nor Spencer, nor Bain, calls it the connection of cause and effect, but only that of antecedent and consequent, how can the connection between matter and thought in the nebula be so clear, that Tyndall can discern in it, at that distance, " the promise and potency of every form and quality of life"? How is it that the relations of matter and mind are unthinkable as they exist in the brain, and thinkable as they exist in the nebula? How is it that the nervous vibrations and the corresponding events of consciousness are, as Tyndall believes them to be, simply consecutive, or correlative,—a case of "parallelism without contact,"—while the matter of the universe, and the life and thought existing in the universe, are so far from being a case of parallelism without contact, that the "potency" of the latter is all in the former?

5. The established definition of matter will, and Tyndall's will not, bear Tyndall's own test of clear mental presentation.

Bishop Butler shows this well enough, even when Tyndall himself, in the Belfast Address, composes the Bishop's argument. Undoubtedly Tyndall has not laid too much emphasis on the famous German saying, "The true is the clear." But his definition, contemplated with all patience and candour, is clear in neither its affirmations nor its negations; while the established is capable of a coherent presentation in both these respects. So far, indeed, is the Belfast Address from knowing its own opinion, that in one place it says the very existence of matter as a reality outside of the mind is "not a fact, but an inference," thus implying that Tyndall is not sure but that Fichte's idealism may be the truth.

6. The established definition is justified, and Tyndall's is not, by the irresistible testimony of consciousness that the will has efficiency as a cause.

Dr. W. B. Carpenter, a far better physiologist than Tyndall, and whose work on "Mental Physiology," just issued, is, always excepting Lotze's "Mikrokosmus," the best discussion produced in modern times of the connection between body and mind, analyses elaborately all the latest facts, including Professor Ferrier's proof of the localisation of functions in the brain; but he saves himself, as Lotze does, from fatalism, materialism, hylozoism, and from that definition of matter which Tyndall adopts. He affirms a very broad and sometimes startling doctrine of unconscious cerebration, but finds in the properties of the nervous mechanism no explanation whatever of our consciousness, that, by acts of will, we can originate physical movements, and control the direction of courses of thought. *The central part of Tyndall's errors is to be found in his shy treatment of this necessary belief.* There results from this shyness his insufficiently clear idea of what he means by causation. Almost while Tyndall was speaking before the British Association at Belfast on atoms, M.

---

[1] Fragments of Science, pp. 120, 121.

Wurtz, president of the French Association, was discussing before that body the same theme, and closing an opening address with no unscientific indistinctness as to what cause signifies. "It is in vain," he said, "that science has revealed to it the structure of the world and the order of all the phenomena: it wishes to mount higher; and in the conviction that things have not in themselves their own *raison d'être*, their support and their origin, it is led to subject them to a first cause,—unique and universal God."[1]

So much does Tyndall's Address lean on Professor Draper's book, on "The Intellectual Development of Europe," that it is a witticism of the London press, that the discourse is rather vapoury when stripped of its drapery; but Draper himself, in an elaborate chapter of his "Human Physiology" (pp. 283-290), undertakes, by an argument on the absolute inertness of nerve arcs and cells in themselves considered, to demonstrate physiologically the existence, independence, immateriality, and immortality, of the soul.

7. The established definition is supported, and Tyndall's is not, by the intuitive belief of the mind as to personal identity.

All the particles of the body are changed within seven years, as science used to teach, or within one year, as it now teaches; and, trite as the power of this objection to materialism has made the objection itself, the inquiry is now more pertinent than ever, How is it thinkable, if matter evolves the personality, that this remains the same, while the physical man does not retain its identity during any two circuits of the seasons?

Mysterious, indeed, is the phenomena of the persistence of physical scars in living flesh that is constantly changing its composition. But grant that the physical basis of memory is an infinite number of infinitesimally small brain-scars, constantly reproduced, although the particles of the brain are all changed, still it is as unthinkable that these scars should rebuild themselves as that the original cuts should cut themselves. It is the generally-accepted theory of metaphysical science, that the soul builds the body, and not the body the soul. But if it be assumed that matter does evolve spirit, then, in the case of the physical basis of memory, it must be supposed to be hand, chisel, inscription, and marble all at once, and not only so, but the reader of the inscription; and all this while every particle of the marble is known to crumble away, and to be replaced by entirely new particles, every twelve months. Flatter contradiction to that principle of the inductive method which asserts that every change must have an adequate cause does not exist anywhere than inheres in all attempts hitherto made to evolve from matter the soul's ineradicable conviction of personal identity.

According to Tyndall's proposed definition, there is in man, as in the universe, but one substance: in the microcosmus, as in the macrocosmus, all is double-faced matter,—spiritual on the one side, and physical on the other. There is nowhere any immaterial agent separate from a material substance. The particles of man's

---

[1] Address republished in "Nature," Aug. 27, 1874.

body are endowed with physical and spiritual properties, and are so peculiarly grouped, that their interaction produces not only his organisation, but his inmost spiritual nature. To say, however, that although the body in its living state loses all its particles, and although these are replaced by new, the old *form* is yet retained, and that this similar grouping of the particles explains the continuity of the consciousness implied in the sense of personal identity, is to introduce design without a designer. *Collocation of parts in an organism is precisely what materialism has never yet explained.* Undoubtedly oxygen and hydrogen have such properties, that, if four atoms of the former and eight of the latter come into proper collocation with each other, they will unite, and form water; but they have no properties tending to bring them together in precisely these proportions. Collocation has ever been a word of evil omen to the materialistic theory.

The particles that go out of the system do not transmit their spiritual any more than their physical qualities to the new particles that come in; for the spiritual qualities, as the changed definition of matter states, inhere in the very substance of each particle; *and inherent properties are not transferable.* When, therefore, we exhale and perspire wasted particles, there is plainly no room left by this definition for denying that we perspire latent soul, and exhale latent personality. In a complete renewal of the particles of the organisation, therefore, there ought to be a renewal of the personality. Such is the theory; but right athwart the only course it can sail in juts up the gnarled rock of man's necessary belief that he does not change his personality: a reef, this, with its roots in the core of the world; a huge, hungry sea-crag, strewn already with the wrecks of multitudes of materialistic fleets, and where the new materialistic Armada is itself destined to beach on chaos.

8. The established definition is justified; and Tyndall's is not by the notorious failure of science to produce a single instance of spontaneous generation.

9. Admissions of the opponents of the established definition exist in abundance to prove, that, if taken in connection with the hypothesis of a creative personal First Cause, it explains all the facts which physical science presents; but these same opponents admit that their definition, even when the doctrine of evolution is accepted, brings the physical inquirer at the end of every possible path of investigation always face to face with insoluble mystery.

10. Finally, the mystic and transcendental definition, by making matter a double somewhat, possessed on its physical side of the qualities claimed for it by established science, but on its spiritual side of the properties necessary to evolve organisation and life, attributes to matter self-contradictory qualities, and is itself inherently self-contradictory.

Matter has extension, impenetrability, figure, divisibility, inertia, colour. Mind has neither. Not one of these terms has any conceivable meaning in application to thought or emotion. What is the shape of love? How many inches long is fear? What is the colour of memory? Since Aristotle and St. Augustine, the antithesis between mind and matter has been held to be so broad, that Sir William Hamilton's common measure for it was the phrase, "the whole diameter of being." *But it is proposed now*—and this is the chief thing I have to say—*to adopt a definition of matter which shall make extension and its absence, inertia and its absence, impenetrability and its absence, divisibility and its absence, form and its absence, colour and its absence, co-inhere in the same substratum.* To this monstrous self-contradiction the mystic hylozoism of Bain, Huxley, and Tyndall, inevitably leads when it defines matter as a double-faced unity, physical on the one side, and spiritual on the other. The reply to this transcendentalism of the evolution school is simply the first law of the syllogistic process, A is not Not-A.

1. Matter and mind have two sets of qualities, each the reverse of the other, and absolutely incapable of co-existence in the same substance.
2. We know that the two sets of qualities exist.
3. We know, therefore, that there are two substances in which the qualities inhere.
4. There is, therefore, a separate immaterial substance.

As to practical inferences from this discussion, it is worth while to note that—

1. The new philosophy as to matter is consistent with a belief in the Divine existence, but not with that of the immortality of the soul. Alexander Bain thinks it absurd to talk of the freedom of the will. Häckel teaches that the will is never free.[1]
2. Teachers of the inductive sciences must not be allowed to play fast and loose with the axioms which lie at the basis of the inductive method. Physics scorning metaphysics is the stream scorning its source. Science, of course, is not science, unless it is inductive. But behind the inductive sciences is an inductive method; and behind the inductive method are the laws of thought. Inductive science implies inductive method; inductive method implies syllogism; syllogism implies axioms; axioms imply intuitive beliefs. Of necessity resting on metaphysics, science has nothing surer than its axioms of intuitive truth; but on precisely those axioms rest the inferences of free-will, responsibility, and the existence of a personal First Cause. Plaintively wrote Aristotle, after mentioning self-evidence, necessity, and universality as the traits of intuitive truth, that they who reject the testimony of the intuitions will find nothing surer on which to build.
3. A distinct definition of the word *natural* ought to put, and ultimately will put, all science on its knees before a personal God. Charles Darwin and Bishop Butler define this fundamental term in the same way; and that not the obscure, heedless, misleading, outworn, and fathomlessly vexatious way common in our brilliant periodical literature. It is a fact in which much solace for timid Christians, and much taming anodyne for audacious small philo-

---

[1] History of Creation, vol. i. p. 237.

sophers, lie capsulate, that the foremost naturalist of our times, and the greatest modern Christian apologist, explicitly agree in affirming—

(1.) That "the only distinct meaning of the word natural" is *stated, fixed, or settled;*" and,

(2.) That "what is natural as much requires and presupposes an intelligent mind to render it so—that is, to effect it continually or at stated times—as what is supernatural or miraculous does to effect it for once."

These far-reaching propositions consist wholly of celebrated words from Butler's "Analogy" (part i. chap. 1), the book which Edmund Burke used to recommend to the acutest of his friends as a cure for scepticism. Barry, the artist, for whose varied and inveterate spiritual sickness Burke prescribed only the study of this volume, was so much benefited by it, that, when he made a painting of Elysium, he placed Butler in the foreground. In our haughty day this renowned passage has become in a new degree famous by being adopted through many editions as the postulate motto on the title-page of Darwin's "Origin of Species." It stands there as a headlight. The agreement of Darwin and Butler as to the meaning of the word *natural* is a beacon which ought to be kept steadily in view by any who grow dizzy as they float, perhaps anchorless, in the surges of modern speculation. Butler's and Darwin's definition is Aristotle's and Kant's and Hamilton's, and Newton's and Cuvier's and Humboldt's, and Faraday's and Dana's and Agassiz's. Just this definition has for ages been the established one in religious science. Of late, as if it were a new discovery, it has appeared as the inspiration of the loftiest portions of modern literature. The vision of what lies behind natural law constitutes the hushed "open secret," which throws the Goethes and Richters, and Carlyles and Brownings, and Tennysons and Emersons, and ought to throw the whole world, into a trance.

4. A miracle is unusual, natural law is habitual, Divine action. The natural is a prolonged and so unnoticed supernatural.

Professor Asa Gray maintains that Charles Darwin is guiltless of all atheistic intent ; that he never denied the possibility of creative intervention in the origin of species ; that he never depended exclusively on natural selection for the explanation of variations in animal forms ; and that he never sneered at the argument from design, to which John Stuart Mill advises philosophers to adhere in their proof of the Divine Existence.

If religion and science are once agreed in adopting Darwin's and Butler's meaning of the word *natural*, all that either of them has to do is to become, in Coleridge's phrase, intoxicated with God.

5. It follows, however, as a minor result of this definition, that it cannot be dangerous to religion to inquire whether the origin of species is attributable wholly to natural causes ; that is, to habitual Divine action. Is it a terrifying thing to ask whether life itself and all its modifications originated in unusual Divine action, or in habi-

tual Divine action, or partly in one, and partly in the other? It is difficult, and to me impossible, to see what ground for disquietude religious science has in the prospect that either of these propositions may obtain proof. What harm, we may say with Charles Kingsley, can come to religion, even if it be demonstrated, not only that God is so wise that He can make all things, but that He is so much wiser than even that, that He can make all things make themselves?

The distinction between mind and matter stands like a reef in the tumbling seas of philosophy; and its roots take hold on the core of the world. In matter there are definite qualities, such as weight, colour, extension. In mind there are none of these : it is absurd to speak of the length of an idea, the colour of a choice, the weight of an emotion. When Tyndall and Bain, and other revivers of the Lucretian materialism, attempt to make the qualities of matter and mind, which differ as diametrical opposites, and by the whole diameter of existence,—extension and the absence of extension, colour and the absence of colour, weight and the absence of weight, inertia and the absence of inertia,—co-inhere in one substratum, and talk of a doublefaced somewhat, " physical on the one side, and spiritual on the other," they are self-contradictory. It is upon the hungry tusks of self-contradiction that whole Armadas of materialistic fleets have been wrecked age after age; and here Tyndall's barge of the gods, which, like Cleopatra's,

" Burned on the water : the poop was beaten gold,
Purple the sails, and so perfumed, that
The winds were love-sick with them,"

only yesterday sank among the mists. But until this reef is exploded, until the distinction between matter and mind is given up, there will very evidently be adequate proof of Design in creation.

Daniel Webster, when once asked if his political opinions on an important topic had changed, wrote to his questioner to look toward Bunker Hill in the morning, and notice whether, in the night, the monument had walked into the sea. If any do not care to puzzle themselves with either the shrill and shallow, or with the more quiet and profound voices of modern speculation, and yet wish freedom from mental unrest, let them not take alarm as to the argument from design until the Aristotelian and age-long monumental distinction between matter and mind has moved from its base ; for, until that shaft walks into the sea, Theism is logically safe. "If," says Kingsley, " there has been an evolution, there must have been an Evolver." " Faith in an order, which is the basis of science," says Asa Gray, " cannot reasonably be separated from faith in an Ordainer, which is the basis of religion." The law of development explains much, but not itself.

6. As science progresses, it draws nearer, in all its forms, to the proof of the Spiritual Origin of Force; that is, of the Divine Immanence in natural law ; that is, of the Omnipresence of a personal First Cause ;

and the religious value of this proof is transcendently great. Wherever science finds heat, light, electricity, it infers the motion of the ultimate particles of matter as the cause ; wherever it finds motion of the ultimate particles of matter, it infers force as the cause ; and, wherever it finds force, it infers, or will yet infer, SPIRIT.

> " God is law, say the wise, O soul, and let us rejoice ;
> For, if He thunder by law, the thunder is yet His voice.
> Speak to Him thou, for He hears, and Spirit with Spirit may meet:
> Closer is He than breathing, and nearer than hands and feet."
> TENNYSON.

II.

THE CONCESSIONS OF EVOLUTIONISTS.[1]

"If everything is governed by law, and if all the power is in the physical universe that ever was there, where is God? In the intention."—Professor BENJAMIN PIERCE, *Unitarian Review*, June 1877, p. 665.

"In regard to the physical universe, it might be better to substitute for the phrase 'government *by* laws' 'government *according to* laws,' meaning thereby the direct exertion of the Divine Will, or operation of the First Cause in the Forces of Nature, according to certain constant uniformities which are simply unchangeable, because, having been originally the expression of Infinite Wisdom, any change would be for the worse."—Dr. W. B. CARPENTER, *Mental Physiology*, chap. xx.

ARISTOTLE said of Socrates that he invented the arts of definition and induction. But Socrates, we know, was not a teacher of logic; he was the investigator of ethical truth; and it was in the endeavour to satisfy a distinctively theological thirst that he smote the rocks at the foot of the Acropolis, and caused to gush forth there these crystalline headsprings of the scientific method. Unless we think boldly, north, south, east, and west, and syllogistically, and on our knees, we do not think at all. A Greek teacher of morals first taught us to think in this manner, and, as instruments of ethical research, invented definition and induction. The scientific method thus had a theological origin. Plato first elaborated it; but he drew all the quenching power of the stream of his philosophy from those pristine springs of definition and induction which Socrates opened. Aristotle, no doubt, was the earliest to give a scientific form to logic as a system; but his river of philosophy was only the continuation of the stream beginning under the Acropolis, where the terrific force of the blow of Socrates had caused these healing waters to burst out. It was in theology that the scientific method first found full application. However much we may criticise the Greek and Latin schoolmen and early theologians, it remains true that they elaborated Aristotle's logic, and drew out of it a system of induction and deduction, which was only turned a little aside to new objects by Bacon. I am not one of those who think Macaulay's essay on Bacon faultless. Gladstone has lately shown that the contrast between the system of

---

[1] The forty-seventh lecture in the Boston Monday Lectureship, delivered in the Meionaon

Aristotle and that of Bacon was not as great as the brilliant historian, who loved antithetical contrasts so well, would make it out to be. The scientific method existed before Bacon's time, and it had received its elaboration chiefly in the schools of theology. But now, since Bacon's time, we hear the scientific method spoken of as if it never had a mother. We are told that religious science must borrow from physical science the scientific method. Religious science will not borrow what is her own. Aristotle affirms that it was in the search after moral truth that Socrates discovered definition and induction. Theology demands in this age, what she has demanded in every age, that we should be loyal to the scientific method. We must have definition; we must have induction; clear ideas and spiritual purposes conjoined are the only deadly intellectual weapons. When a haughty attitude is assumed by physical science in the name of the scientific method, all that religious science has to do is to show that she was the mother of that method, to adhere to it herself, and to hold to it, a little mercilessly, physical science also.

Among the concessions of evolutionists, these are notorious :—

1. That spontaneous generation must have occurred, or the doctrine of evolution as held by Huxley and his school cannot be true.

2. That spontaneous generation has never been known to occur.

3. That it is against all the ascertained analogy of nature to suppose that it ever has occurred.

4. That, if spontaneous generation has not occurred, it must be admitted that a supernatural act originated life in the primordial cell or cells.

5. That the doctrine of evolution as held by Huxley cannot be true, unless some bridge can be found to span the chasm between the living and the not-living.

6. That the present state of knowledge furnishes us with no such bridge.

Who makes all these far-reaching concessions? Professor Huxley. Where? In a most suggestive article on "Biology," published in "The Encyclopædia Britannica," the ninth edition of which, as you are aware, is now issuing from the press.

It is not asserted by this Lectureship that a doctrine of natural selection cannot be proved unless spontaneous generation can be shown to be a possibility. I assert, however, that the doctrine of evolution, "as held by Huxley and his school," cannot stand, unless spontaneous generation can be shown to have been a fact. This is Huxley's own concession. He says, "*If the hypothesis of evolution is true, living matter must have arisen from not-living matter;* for by the hypothesis the condition of the globe was at one time such, that living matter could not have existed in it, life being entirely incompatible with the gaseous state." [1]

"The properties of living matter distinguish it absolutely from all other kinds of things; and *the present state of knowledge furnishes us with no link between the living and the not-living.*" [2]

---

[1] Professor T. H. Huxley, Encyc. Brit., ed of 1876, art. Biology, p. 689. [2] Ibid., p. 679.

"At the present moment there is not a shadow of trustworthy direct evidence that abiogenesis [or spontaneous generation] does take place, or has taken place, within the period during which the existence of the globe is recorded."[1]

Will you put these strategic propositions into contact with each other? Huxley's form of the doctrine of evolution stands or falls with the fate of the doctrine concerning spontaneous generation. Darwin's form of it does not ; Dana's not ; and Gray's not.

Huxley, you notice, expressly concedes that all the evidence we now have is against the theory that spontaneous generation is possible, and that the present state of knowledge furnishes us with no link between the not-living and the living.

Häckel concedes, and it is very evident from the nature of the case, that if the primordial cells did not originate spontaneously, or by usual Divine action, they must have been originated supernaturally, or by unusual Divine action. The theory of natural selection as held by Darwin does not attempt to bridge the chasm between the living and the not-living.

To show how incisive the assertion is, "that life is incompatible with the gaseous state," Professor Huxley says, in a note following the sentence I have read, that it makes no difference, if we adopt Sir William Thomson's theory, that life may have been inducted into this planet from life in some exterior physical source. The nebular hypothesis, which is a part of the great evolution theory, asserts that all the worlds were once in a gaseous state ; and so in that exterior physical source, which was once a gas, how could life have arisen? Even Tyndall's famous matter, so richly endowed as to have in it "the potency and promise of all life," must itself once have been in a gaseous state.

When Professor Huxley and Professor Tyndall sit together at the top of the Alps, and Tyndall begins his definition of matter, if Professor Huxley will whisper to him these words, "that life is entirely incompatible with the gaseous state," it will not be logically competent to Professor Tyndall to go on speculating, as he once did on the Matterhorn, whether or not his pensiveness and his thoughtfulness, as well as the gnarled granite peaks, were all potentially existent in the earliest nebula. Let Professor Huxley and Professor Tyndall correct each other, and perhaps there may arise, in that way, contagious life by collision.

"But," continues Professor Huxley, "living matter once originated, there is no necessity for another origination, since the hypothesis postulates the unlimited, though perhaps not indefinite, modifiability of such matter. Of the causes which have led to the origination of living matter, it may be said that we know absolutely nothing."

Here is determined agnosticism. Of course, if physicists will not look outside of matter, they can have no knowledge of a first cause. "Give me matter," said Kant, "and I will explain the formation of

---

[1] Professor T. H. Huxley, Encyc. Brit., ed. of 1876, art. Biology. p. 689.

a world; but give me matter only, and I cannot explain the formation of a caterpillar." Professor Huxley likes to quote the first half of that celebrated saying, without the last.

To test the value of these concessions by Huxley as to spontaneous generation, take another theme, and one on which our opinions are not divided—the philosopher's stone. We do not now find ourselves able to make a philosopher's stone. We have no reason to believe that Nature ever made a stone that will transmute the baser metals into gold. There is nothing in science to show that such a stone can be found or made. But, unless such a stone has been made at some time in the past, we must give up a pet theory in philosophy. Therefore let us assert, that, in the complex conditions of a cooling planet, perhaps the philosopher's stone may have come into existence by fortuitous concourse of atoms. You smile, gentlemen, because you are true to the scientific method, and I mean you shall be. But Strauss, in his "Old Faith and New," asks, "Who can tell what may have occurred in a cooling planet?" Virchow says that things were mixed in those early ages, and that it must be that somehow life originated spontaneously; at least Strauss would be very glad to have us prove a negative.

Now, gentlemen, there is a famous theory in geology called the Uniformitarian Hypothesis. It assumes that the geological formation of the globe was due to precisely the same physical forces that now exist. We have given up the idea of great catastrophes in geology. But when we reason concerning spontaneous generation, if we take our stand on the further side of the fact—if it ever was a fact,—we are in the field of simple physical forces. Here are just the influences that brought into existence our mountains and seas, and determined events in the inorganic world. According to all established science, these forces have been uniform. The Uniformitarian Hypothesis turns upon the idea that uniformity exists in the forces of the inorganic world. We must, therefore, insist, that, if spontaneous generation does not occur now, it never occurred. We must do this in the name of the uniformity of nature.

The chasm between the not-living and the living forms of matter is the fathomless abyss at the ragged edge of which every traveller on atheistic or agnostic roads at last lifts his foot over thin air.

It is notorious that evolutionists admit,—

7. That natural selection cannot have originated species, if the sterility of hybrids is a fact.

8. That, in the present state of knowledge, the sterility of hybrids must be accepted as a fact.

9. That it is fair to ask, as a proof of evolution, that there be formed by selective breeding two species so different that their intercourse will produce sterile hybrids.

10. That no such species have as yet been formed by selective breeding, and that, until two such have been formed, the strongest proof of the doctrine of evolution is wanting.

Who admits all this? Professor Huxley. Where? In his famous "Lay Sermons and Reviews," where he cites (p. 308, American edition) Professor Külliker, than whom there is no greater authority in embryology. This German says, "Great weight must be attached to the objection brought forward by Huxley, otherwise a warm supporter of Darwin's hypothesis, that we know of no varieties which are sterile with one another, as is the rule among sharply distinguished animal forms. If Darwin is right, it must be demonstrated that forms may be produced by selection, which, like the present sharply distinguished animal forms, are infertile when coupled with one another; and this has not been done."

What, now, does Professor Huxley himself say, speaking before scholars, and in reply to this passage? "The weight of this objection is obvious," is his answer; "but our ignorance of the conditions of fertility and sterility,"—which have been witnessed by man six thousand years, at least,—"the want of careful experiments extending over a long series of years, and the strange anomalies presented by the cross fertilisation of many plants, should all, as Mr. Darwin has urged, be *taken into account* in considering it." This is all he says, or that can be said, in reply to this objection.

Häckel asserts that sometimes hybrids are not, and five hundred other authorities, and all the proverbs of breeders, affirm that true hybrids are, sterile.

It is safe to say that evolutionists concede,—

11. That natural selection cannot take leaps, and that therefore a multitude of links must have existed between man and the higher apes.

12. That after a diligent search, for nearly forty years, for traces of these missing links, none have been found.

13. That, in spite of all imperfections of the geological record, the destruction of these relics, without traces, is amazing, and that their absence leaves the argument for evolution weakest where it should be strongest.

14. That the oldest human fossils exhibit in essential characteristics no approach to the ape type.

"No remains of fossil man," says Professor Dana, in a most significant passage of his "Geology" (edition of 1875, p. 603), "bear evidence to less perfect erectness of structure than in civilised man, or to any nearer approach to the man-ape in essential characteristics. The existing man-apes belong to lines that reached up to them as their ultimatum; but, of that line which is supposed to have reached upward to man, not the first link below the lowest level of existing man has yet been found. This is the more extraordinary, in view of the fact, that, from the lowest limits in existing man, there are all possible gradations up to the highest; while below that limit there is an abrupt fall to the ape-level, in which the cubic capacity of the brain is one half less. *If the links ever existed, their annihilation without trace is so extremely improbable, that it may be pronounced impossible. Until some are found, science cannot assert that they ever existed.*"

In regard to these missing links, Darwin himself says that their absence is amazing. Even Huxley says of what is unquestionably one of the oldest fossil skeletons of man, that it has "a fair, average human skull." The lengths of the bones of the arm and thigh of the man of Mentone, one of the oldest human fossils yet discovered, have the proportions ordinarily found in man, and the skull is of excellent Caucasian type.[1] The poorest fossil human brain is twice the cubic capacity of the best ape brain.[2]

It must be noticed that evolutionists admit,—

15. That, if any animal can be shown to possess organs or peculiarities of no use to it in the struggle for existence, the theory of natural selection breaks down.

16. That the hairlessness of man was not only of no use, but was a disadvantage, to him in the struggle for existence, and cannot be accounted for by natural selection, and must be accounted for by sexual selection.

17. That many animals possess peculiarities which, so far as we can see, can be of no use to them in the struggle for existence, and cannot be accounted for by any form of selection, natural or sexual.

In his "Descent of Man," published in 1871, Mr. Darwin himself makes these great concessions. "Natural selection," said Mr. Darwin in his "Origin of Species," published in 1859, "can act only by taking advantage of slight successive variations ; it can never take a leap, but must advance by short and slow stages. If it could be demonstrated that any complex organ existed which could not possibly have been formed by numerous successive slight modifications, my theory would absolutely break down."

Compare that extract with this : " I now admit, after reading the essay of Nägeli on plants, and the remarks by various authors with respect to animals, that, in the earlier editions of my 'Origin of Species,' *I probably attributed too much to the action of natural selection or the survival of the fittest. I had not formerly sufficiently considered the existence of many structures which appear to be, as far as we can judge, neither beneficial nor injurious ;* and this I believe to be one of the greatest oversights as yet detected in my works."[3]

It may be safely asserted that evolutionists concede,—

18. That whether the cause of variation is a force exterior or one interior to the modified organism, or a combination of these forces, is not known.

19. That it is probable that variation is due much more to some innate force in the modified organism than to anything outside of it.

20. That the influence of natural selection has been exaggerated ; that it explains much, but not everything ; that it deserves only a co-ordinate rank with sexual selection as the explanation of the origin of man ; and that very possibly it should have a subordinate rank in contrast with yet unknown causes of variation.

"*No doubt man, as well as every other animal*," says the Charles Darwin of to-day, "*presents structures which, as far as we can judge with our little knowledge, are not now of any service to him, nor have*

---

[1] See Dana's Geology, frontispiece, and pp. 575, 577, and 603.
[2] Ibid., 603.   [3] Descent of Man, English edition, vol. i. p. 152.

been so during any former period of his existence, either in relation to his general conditions of life, or of one sex to the other. Such structures cannot be accounted for by any form of selection, or by the inherited effects of the use and disuse of parts." [1] "In the greater number of cases we can only say that *the cause of each slight variation and of each monstrosity lies much more in the nature or constitution of the organism than in the nature of the surrounding conditions,* though new and changed conditions certainly play an important part in exciting organic changes of all kinds." [2]

These astonishing modifications of his own theory by Darwin induce Professor St. George Mivart to assert in his "Lessons from Nature," a work which has but just crossed the Atlantic, that "the hypothesis of natural selection originally put forward as the origin of species has been really abandoned by Mr. Darwin himself, and is untenable. It is a misleading positive term, denoting negative effects, and, as made use of by those who would attribute to it the origin of man, is an irrational conception,"—"a puerile hypothesis." [3] Any who remember Professor Huxley's article on Darwin's Critics, in "The Contemporary Review," for November 1871, will recall the strong terms in which he speaks of Mivart's scientific and philosophical competence. But Mivart holds nearly Professor Theophilus Parson's and Owen's creed, that species have originated by a force interior, and not exterior, to the modified organism. *To that position Darwin draws nearer and nearer.* Among Darwinians there seems to be a conspiracy of silence as to this fact. Darwinism is becoming Owenism. Darwin himself is not a good Darwinian.

God be thanked that this age takes nothing for granted! No: it does take one thing for granted,—its own superiority to all other ages; and yet one other thing,—that there are *not* more things in heaven and earth than are dreamed of in its philosophy. But, my friends, the scientific method requires, that, when we run up our list of causes,—chemical, electrical, physical, mental, spiritual,—we should put at the top, to reach on into the infinite, another class,—the unknown. Even in the nineteenth century, there are more things in heaven and earth than are dreamed of in our philosophy.

---

[1] Descent of Man, vol. ii. p. 387.   [2] Ibid., vol. ii. p. 388.
[3] Professor St. George Mivart, Lessons from Nature, London, 1876, pp. 280-331.

III.
THE CONCESSIONS OF EVOLUTIONISTS.[1]

"The convertibility of the physical forces, the correlation of these with the vital, and the intimacy of that *nexus* between mental and bodily activity, which, explain it as we may, cannot be denied, all lead upward towards one and the same conclusion,—the source of all Power in mind; and that philosophical conclusion is the apex of a pyramid, which has its foundation in the primitive instincts of humanity."—Dr. W. B. CARPENTER, *Mental Physiology*, chap. xx.

"Causation is the Will, Creation the Act, of God."—W. R. GROVE, *Essay on the Correlation of Physical Forces.*

THE small philosopher is a great character in New England. His fundamental rule of logical procedure is to guess at the half, and multiply by two. God be thanked for the diffusion of knowledge! God save us from the attendant temporary evils of arrogant sciolism in democratic ages! These are a necessary transitory stage in the progress of popular enlightenment which has just begun to dawn in this yet dim Western world. A little knowledge is a dangerous thing; and it is our boast that, in America, every man has a little knowledge. We must drink deep, or taste not the Pierian spring; but every breathlessly hurried free citizen now is endeavouring, to his honour, to have a taste at least; and yet we know how mercilessly commerce and greed, and the toil for daily bread, wrench parched lips away from the deep draught. Full popular enlightenment is popular sanity; penumbral popular enlightenment is often popular insanity; and yet the penumbral must precede the full radiance. The small philosopher is always a great character under representative institutions. He seems destined to reign long on the earth, and often disastrously, and yet not for ever. We are an atrociously independent and as yet only a half-educated people. De Tocqueville said that individualism is the natural, and must often be a most mischievous, basis of democratic philosophy. To her great credit and to her great temporary mental distress, Massachusetts, in which popular enlightenment is more widely diffused than elsewhere, has probably just now more small philosophers than any other population of equal size on the globe. Emerson wrote of average Massa-

[1] The forty-eighth lecture in the Boston Monday Lectureship, delivered in the Meionaon

chusetts as she was thirty years ago, "It is a whole population of ladies and gentlemen out in search of a religion." No doubt it is to our credit that we study the newspapers; but it is not to our credit that we do not better maintain the best ones, and that we do not sift newspaper information a little more warily, and that some of us think a man can be competently educated on the most trustworthy part of the daily press. "We must destroy the faith of the people in the penny newspaper," I once heard Carlyle say in his study at Chelsea. I fathomlessly respect able and conscientious newspapers; I revere their majestic mission in history. I used to be told in Europe that Americans are governed by newspapers; and I was accustomed to answer, "No, gentlemen, not by newspapers, but by news—a very different thing." But, whether the shrewdest readers get at the news that is the most strategic in science, in politics, in art, in theology, by a hasty scramble through the midnight scribble of our cheaper dailies, is rather doubtful, or, rather, not doubtful at all. The most appropriate prayer, when one takes up the penny newspaper, is an invocation of the spirit of unbelief. But the best-used book of your small philosopher is the newspaper. He is unchurched in art, in science, in theology. He hears great names; he obtains glimpses of great truths; he puts half-truths in the place of systems that will bear the microscope; and when religious science occasionally gets his haughty hearing, it cannot on the Sabbath-day go into secular discussion with him, and you cannot hold his attention at first, except by secular discussion. You say that I am using this Lectureship very maladroitly, and that it is not wise to discuss here evolution and materialism. I do not speak to or for ministers or scholars, although they crowd this hall; I am talking to small philosophers.

Lord Bacon said that "truth emerges sooner from error than from confusion;" and, in the spirit of that remark, you will allow me to be analytical, and to number my propositions, in order that I may save time, and yet be distinct in a crowded discussion. Twenty concessions having been mentioned in a previous lecture, it is next to be noticed that it is notorious that evolutionists admit,—

21. That life is incompatible with the gaseous state, or the state of fused metals.

22. That our present knowledge justifies the conclusion, that probably two hundred millions, and certainly five hundred millions, of years ago, the earth and the sun were in a fused state.

23. That neither two hundred nor five hundred millions of years are enough to account for the formation of plants and animals from primordial cells on the theory of the Darwinian transmutation.

These, gentlemen, are the outlines of what many men of science regard as the most serious of all objections to the hypothesis of evolution. This is the only difficulty to which Professor Huxley in his New York lectures condescended to reply, it is the most prominent of the objections which Häckel endeavours to refute in his recent daring work on "The History of Creation." I now hold in my hand this book, of which Darwin himself says, that its author has much more information than he has on many points, and that, if it had appeared before "The Descent of Man," the latter work would probably never have been written. Professor Häckel teaches at present in the University of Jena, in Germany ; and he is one of the most extreme of evolutionists. He denies the freedom of the will, and is a thorough-going defender of the theory of the possibility of spontaneous generation.[1] He affirms, as Huxley does, that we have no direct evidence that spontaneons generation has ever occurred, and that it is against all the analogy of current nature to suppose that it has occurred. But he knows the exigencies of the radical form of the theory of evolution ; and so he assumes, with Strauss, that possibly in a cooling planet a living cell may have been originated by the fortuitous concourse of atoms. A cell once originated, we can account for all life. But he is painfully aware that the Darwinian transmutation requires almost immeasurable time. " In the same way," he says, "as the distances between the different planetary systems are not calculated by miles, but by *Sirius-distances*, *each of which* comprises millions of miles, so the organic history of the earth must not be calculated by thousands of years, but by paleontological or geological periods, *each of which* comprises many thousands of years, and perhaps millions, or even milliards of thousands of years."[2] To the same effect speak Lyell and Dana, and even Darwin.[3]

Now, Professor Huxley very strangely said, in his lectures in New York, that, if the astronomer and geologist will settle between themselves the question as to the length of geological time, he will " agree with *any* conclusion."

Not so speaks the candid Darwin ; not so the audacious Häckel ; not so Lyell ; not so Dana ; not so any cautious evolutionist ; not so even Huxley himself, when he talks before scholars.

" Thousands of millions of years," says Dana,[4] " have been claimed by some geologists for time since life began. Sir William Thomson has reduced the estimate, on physical grounds, to one hundred millions of years as a maximum." " Any " conclusion ! Let us take the best estimate there is, that of one hundred million years ; and Häckel implicitly affirms that this is not enough for the process of the Darwinian transmutation.

What is the evidence, gentlemen, that our earth and the sun were in a molten condition, say, five hundred millions of years ago ? We tolerably well know of what materials the sun is composed. We bring down by the spectroscope its talkative rays, and we can tell what metals are in it. We know the nature of these metals on our globe. Heat is the same thing here and there ; gravitation, the same here and there ; light, the same here and there. *The immense argument of analogy makes us sure of our footing just so far as the unity of nature prevails.* We can estimate approximately what the

---

[1] Häckel, History of Creation, chap. xiii.   [2] Ibid., chap. xxiv.
[3] Lyell, Geology, vol. i. pp. 234, 235 ; Dana, Geology, ed. of 1875, p. 591 ; Darwin, Origin of Species, p. 286.   [4] Geology, pp. 59, 591.

heat must have been that would fuse the globe and the sun. Sir William Thomson, whose scientific eminence no man will deny, went into a very laboured calculation, not long ago, to determine how many years since it was that the sun was a molten mass, and how many years since it was that the globe was in a fused state; and it is very significant that he came to the same conclusion in both cases. The two conclusions tallied. The sun, he said, must have been in a molten state four hundred millions of years ago at the most; and it probably was in that state two hundred millions of years ago at the least. The same may be said of the earth, which, however, was not cool enough to admit life until about one hundred millions of years ago, as Dana says.

When we look at the reasons why Professor Huxley sneers at this argument, we are the more amazed. "The biologist," he says, "knows *nothing whatever* of the amount of time which may be required for the processes of evolution." Does not he know that there is an immense extent of time required for it ? "Nothing whatever" known about the period needed! Why, all Darwinians are agreed, all evolutionists are agreed, that we must take Sirius-distances to measure the time required by evolution. "I have not *the slightest* means of guessing," said Professor Huxley at New York, "whether it took a million of years, or ten millions, or an hundred millions of years, or a thousand millions of years, to give rise to that series of changes." On Darwin's, Lyell's, Dana's, and Häckel's authority, this must be called careless talk. It leaves a colossal objection unshattered.[1]

It is admitted by evolutionists,—

24. That variability in species is a lessening quantity as descendants are farther and farther removed in form from their progenitors.

25. That, as every lessening must be a finite quantity, species are known to vary only within comparatively narrow limits.

26. That selective breeding has thus far found variability a limited quantity.

27. That the observed differences caused by variability are infinitely small as compared with the range of variability required by the Darwinian theory.

It has been well said that the savage, looking upon a projectile of modern artillery, might carelessly think it would reach the stars. He does not make allowance for the circumstance that the speed of the ball is a lessening quantity. We find it to be a fact, that, the farther a derived animal form is removed from its progenitor, the less and less rapidly variations proceed. It follows, therefore, that these lessening variations may be fitly represented by a sphere, the original progenitor being the centre, from which there may be variations in all directions, and to which there may be reversions in any direction.[2] The variations are like the throwing-up of a cannon ball from the earth; the motion away from the central point is

---

[1] See North British Review, 1867, vol. xlvi. p. 304.
[2] Ibid., art. on "The Origin of Species."

slower and slower as the distance between the ball and the central point is greater and greater. *We assuredly know that it is a truth of science that variability is a lessening quantity; and we therefore do know mathematically that there are limits to variability; for every lessening number is a finite quantity.* Thus, gentlemen, there are broad distinctions to be made between so-called species of a variable and real species of an unvarying kind. If we are to be abreast of our modern science, we shall be shy of saying that there is nothing which has been *called* a species which may be transmuted into another species.

I would confine the definition of species to the limits of ascertained variability. Here is the sphere of variation; and we know that the more any descendant varies from its progenitor, the more likely it is to revert. It may go back in a single generation. The law of science is, that variability, being a lessening, is a finite quantity. If you will draw a circle around the outermost sphere of variability, you will have what Häckel calls a "good species" in distinction from a merely nominal species. The thing we need most in the discussion of evolution is a new definition of species. *A real species will be conterminous with the outermost limits of the sphere of ascertained variability. Grant me this definition, and I will stand with established science on the fact that we have no direct evidence that any real species, thus defined, has ever been transmuted into another species.*

It is notorious that evolutionists concede,—

28. That the cubic capacity of the brain of the highest apes is thirty-four inches, and of the lowest men sixty-eight.

29. That the brain of man is by much larger than he needed in the struggle for existence.

30. That the struggle for existence, or natural selection, does not account for the brain of man.

31. That the eye of the trilobite, one of the oldest of fossil forms, is fully developed and perfect.

32. That the trilobites appear suddenly in the geological record; that there are no premonitions of their approach; and that there is as yet no direct evidence that they had any ancestry.

33. *That the use of an organ may account for its modification, but not for its formation, since it cannot be used until it is formed.*

34. That in many cases, like those of the eye of the trilobite and the brain of man, not only the theory of natural selection, but that of sexual selection, breaks down completely.

35. That in some cases it is impossible to imagine what has produced useful variations in animal forms.

36. That, in certain instances, the adaptation of means to ends cannot be accidental, but must be referred, not to natural, but to supernatural law; that is, not to the habitual, but to unusual divine action.

These, gentlemen, are startling concessions; and the most startling of them all is the last, that there are instances in which the adaptation of means to ends "cannot be accidental." But those are Darwin's words. You will remember that in his delicious book on the "Fertilization of Orchids," at the end of its first chapter he speaks of a marvellous arrangement by which, in one species of these flowers, the sipping-moths are "purposely delayed in obtaining nectar." He says, "If this is accidental, it is a fortunate accident for the plant. If this be not accidental, *and I cannot believe it to be accidental*, what a singular case of adaptation!" Professor Mivart[1] quotes several similar admissions from Darwin's later writings; and he regards them as a virtual, though not explicit, retraction of the theory of natural selection. You say these are all careless expressions on the part of Darwin? I beg pardon: they are not so understood by men of scientific competence, some of whom watch him more closely than the tiger watches its prey.

I am not one of those who lie in wait to find fallacies in Darwin; for it matters little to me, as a student of religious science, which one of the three or four theistic systems of evolution is proven to be the best. If there is a change, I know that every change must have an adequate cause. If there is order in the universe, I know there must have been an Ordainer; for every change must have had an adequate cause. Based upon incontrovertible axiomatic truth, any man may stand in the yeasting seas of speculation, and feel that victorious reef tremorless beneath him; ay, and fall asleep on it, while the rock, in muffled stern thunders, speaks to the waste, howling midnight surge, "Aha! thus far ye come, but no farther." Men can never give up belief in causation. If we know there has been evolution in the universe, we know that there has been an Evolver; and, if design, a Designer; for every change must have a sufficient cause. It will not be to-morrow, nor the day after, that men will give up self-evident, axiomatic truths.

Owen, Parsons, Mivart, Dana, and Darwin himself, all admit that useless characteristics and organs cannot be explained by natural selection; and Darwin has made lately many admissions of his oversights on this point.

Dana, to the latest date, disagrees completely with Huxley and Häckel as to the origin of man, and agrees with Owen, Gray, Mivart, Parsons, and the whole long, stately, and growing list of the theistic school.

It is not denied anywhere, that a certain extent of variation may be experimentally produced by external conditions, as in the brine shrimp and the axiolott. What is denied is, that external conditions can account for the difference between the not-living and the living.

It seems to be the policy of atheistic and agnostic evolutionists to obscure the distinction between *a* theory and *the* theory of evolution. The tendency of science is in favour of the former, and against the latter; that is, for Dana and Hermann Lotze, and against Herbert Spencer and Häckel. The different schools of evolutionists must be distinguished, or there can be no clearness of discussion on this theme.

You will allow me to read one passage from Professor Dana on

---

[1] Lessons from Nature, 1876, chaps. ix. and x.

THE CONCESSIONS OF EVOLUTIONISTS. 31

the great contrast between the brain of man and that of apes. Professor Dana, with respect be it said, is not a Darwinian; it is hardly fair to call him, without qualification, an evolutionist. He believes that evolution explains much; he does not believe that it explains everything. He does not account for man by evolution. He agrees with Wallace, Darwin's great coadjutor, with regard to the origin of the human will and conscience. Professor Dana, in justifying his significant concessions, says,[1] "In the case of man, the abruptness of transition 'from preceding forms' is still more extraordinary, and especially because it occurs so near to the present time. In the highest man-ape, the nearest allied of living species has the capacity of the cranium but thirty-four cubic inches; while the skeleton throughout is not fitted for an erect position, and the fore-limbs are essential to locomotion: but, in the lowest of existing men, the capacity of the cranium is sixty-eight cubic inches; every bone is made and adjusted for the erect position; and the fore-limbs, instead of being required in locomotion, are wholly taken from the ground, and have other and higher uses."

You will be told that Professor Huxley has said that man differs less from the apes than the upper apes do from the lower apes, or than the uppermost men from the lowermost. You will be assured that there is this and that and yet another point of resemblance between the skeletons of man and of the apes. But bring the contrast to the real test. What of the brain? That is the central portion of the system: increased cephalisation is the law of the progress of animal forms; and, the moment you compare man and the ape on that strategic point, the difference is half.

Thirty-four cubic inches of cranial capacity on the animal side, sixty-eight on the human, and no link between the two! Forty years given to the search! All the agony of the defence of the Darwinian hypothesis engaged in all quarters of the globe in filling up this tremendous gap, and the colossal absence yet remaining!

Professor Agassiz lies in Mount Auburn yonder; and on his breast there is a boulder from his native Alps. Whenever I look on it, I think what a boulder that man may have carried on his breast into his grave, because he was not able to develop the proposition which he laid down as a gauntlet before Darwinism in the last article he ever printed. You remember that in our brilliant Atlantic Monthly, face to face with the world, Professor Agassiz, a few days before he passed into that Unseen Holy where all puzzles are solved, affirmed that it can be proved that the geological record is not so imperfect but that we know what existed between the highest apes and the lowest men, and that, however broken it may be, "there is a complete sequence in many parts of it, from which the character of the succession may be *determined*."[2] He promised to prove that. He bent that colossal bow, and it dropped out of his dying hand. On the English-speaking globe, now that Lyell has gone hence, there is no man but Dana that can take up that bow, and bend it. But what does Dana say? Go

---

[1] Geology, p. 603.   [2] Atlantic Monthly, vol. xxxiii. p. 101.

to Agassiz's grave ; take with you those yet moist sheets of the last number of the American Journal of Science and Arts ; read over Agassiz's tomb the latest utterance of the highest and gravest authority in American geological science, and you may bring solace to a hovering, mighty spirit for an unfinished task. You will read Dana's latest words : [1] "*For the development of man, gifted with high reason and will, and thus made a power above Nature, there was required, as Wallace has urged, a special act of a Being above Nature, whose supreme Will is not only the source of natural law, but the working-force of Nature herself. This I still hold.*" You say that Agassiz was unduly theistic, and assumed that there is nothing in evolution. Dana is more cautious. The present state of knowledge, he says,[2] favours the theory that " the evolution of the system of life went forward through the derivation of species from species, according to natural methods not clearly understood, and with few occasions for supernatural intervention. The method of evolution admitted of abrupt transitions between species ; but for the development of man there was required the special act of a being above Nature, whose supreme will is the source of natural law." Huxley has come ; Huxley has spoken ; Huxley has gone ; and Dana, over Agassiz's grave, joins hands with Agassiz in the Unseen Holy, to affirm that man is the breath of God.

It is notorious that evolutionists concede,—

37. That "molecular law is the profoundest expression of the Divine Will." This is Dana's language.[3]

38. That, therefore, even if the nebular hypothesis be accepted, design in creation yet stands proved.

39. That, even if spontaneous generation under molecular law were demonstrated, the fact of design in creation would yet stand proved.

If you will elaborately master Professor Stanley Jevon's famous work on the " Principles of Science," you probably will come to his theistic conclusions, *even if you believe in the possibility of spontaneous generation under molecular law.* We have had important works on the logical method and order, from Aristotle to Kant and Hamilton ; and yet, Professor Pierce of Harvard being judge, there have been few more important productions on that theme than the " Principles of Science," by Stanley Jevons, professor of logic and political economy at Owens's College, Manchester. He is an evolutionist ; but he is also a logician.

"*I cannot,*" he says, "*for a moment admit that the theory of evolution will alter our theological ideas.* . . . The precise reason why we have a backbone, two hands with opposable thumbs, an erect stature, a complex brain, about two hundred and twenty-three bones, and many other peculiarities, is only to be found in the original act of creation. *I do not, any less than Paley, believe that the eye of man manifests design.* I believe that the eye was gradually developed ;

---

[1] American Journal of Science and Arts. October 1876, p. 251.
[2] Geology, pp. 603, 604.   [3] Amer. Jour., October 1876, p. 250.

but the *ultimate result must have been contained in the aggregate of causes; and these so far as we can see, were subject to the arbitrary choice of the Creator.*"[1]

It is notorious that even Tyndall concedes,—

40. That if a right-hand spiral movement of the particles of the brain could be shown to occur in love, and a left-hand spiral movement in hate, we should be as far off as ever from understanding the connection of this physical motion with the spiritual manifestations.[2]

It is conceded by Dana,—

41. That the possession by man of free-will and conscience shows that he must have been brought into existence by a being at least as perfect as himself; that is, by an agency possessing free-will and conscience.

42. That evolutionists are of two schools,—the extravagant and the moderate, or the wholesale and the discriminating; and that the former do, and the latter do not, account for man by the theory of evolution.

Häckel concedes,—

43. That the theory of man's descent from apes is, according to the admission of the wholesale evolutionists, deductive, and not inductive,—a result of speculation, and not of observation.

44. That it probably can never be established by the inductive, that is, by the most strictly scientific method.

Do you suppose that I think that this audience can be cheated? I do not know where in America there is another weekly audience with as many brains in it; at least I do not know where in New England I should be so likely to be tripped up if I were to make an incorrect statement, as here. "The process of deduction," says Häckel, "is not based upon any direct experience. Induction is a logical system of forming conclusions from the special to the general, by which we advance from many individual experiences to a general law. Deduction, on the other hand, draws conclusion from the general to the special, from a general law of nature to an individual case. Thus *the theory of descent is, without doubt,* a great inductive law, empirically based upon all biological experience. *The theory, on the other hand, which asserts that man has developed out of lower, and, in the first place, out of ape-like mammals, is a deductive law inseparably connected with the general inductive law.*"[3]

The theory of man's origin from apes is not based upon direct experience. Merely deductive conclusions from circumstantial evidence are sometimes lawful. We do not know all about the worlds beyond the sweep of the telescope; but so firmly is the theory of gravitation established that we believe that, if a new world should be discovered, it would be found to be under the law of gravitation. *If you will prove by induction the system of evolution as thoroughly as the Copernican system has been proved by induction, you may then fill gaps by deduction.* Astronomers predict sometimes that eclipses will occur, and they do occur according to prediction; and we think,

---

[1] Professor W. Stanley Jevons, Principles of Science, vol. ii. pp. 461, 462.
[2] Fragments of Science, pp. 120, 121. [3] Häckel's History of Creation, vol. ii. p. 357.

therefore, that we have ascertained something conclusive as to the mechanism of the heavens. *If evolutionists can by selective breeding produce from the same stock two varieties so widely differing that their crossing will produce sterile hybrids, then I will say that they have a scientific right to fill up by deduction the gaps in the direct evidences of evolution, and not till then.*

Professor Häckel further concedes, —

45. *That "most naturalists, even at the present day, are inclined to give up the attempt at natural explanation" of the origin of life, "and take refuge in the miracle of inconceivable creation."* [1]

The trouble with your small philosopher in Massachusetts and England is, that he out-Darwins Darwin and out-Häckels Häckel. It is important, at times, that the pulpit should show that it is not afraid of these topics ; and you will notice, that, in this Lectureship, the theme of evolution is not skipped.

You will pardon me one further word on Bathybius, which Professor St. George Mivart calls a sea-mare's nest.

" No more of that, Hal, an thou lovest me."

Häckel has minutely figured Bathybius in the plates of his most elaborate works. Huxley named it from Häckel, *Bathybius Hackelii.* Strauss rested on Bathybius the central arch of his argument against the supernatural.

It was the haughty claim of Huxley and Strauss and Häckel,—

46. That Bathybius is an organism without organs.

47. That it performs the acts of nutrition and propagation.

48. That, with other organisms like itself, it stands at the head of the terrestrial history of the development of life.

49. That it spans the chasm between the living and the not-living.

50. That it renders belief in miracle impossible.

Häckel makes Bathybius a stem from which all terrestrial life divides, and comes to its present state.[2] It would not be worth much for me here to cut down this or that bough in the great tree ; but if, with the latest scientific intelligence, I may strike at its bottom stem, Bathybius, I shall have done something. You must not think that students of religious science have no right to be interested in this classical organism. We have heard of it in theological works. We had it thrust in our faces as proof that a miracle is impossible. We therefore are interested, when, walking past our bookstores, we can pick up the yet fresh sheets of the American Journal of Science and Arts, and turn to a passage on Bathybius in an article on the voyage of the ship " Challenger." Will gentlemen here do themselves the justice, and this topic the justice, to read this authoritative intelligence (October Number, pp. 267, 268) ? You will find there this closing concession :—

51. That Bathybius has been discovered in 1875 by the ship "Challenger" to be—Hear, O heavens! and give ear, O earth!—sulphate of lime ; and that, when dissolved, it crystallises as gypsum.

---

[1] Häckel's History of Creation, vol. i. p. 327.
[2] Ibid., vol. i. pp. 184, 344 and vol. ii. p. 53.

IV.
THE MICROSCOPE AND MATERIALISM.[1]

ὀλιγοδρανέες, πλάσματα πηλοῦ, σκιοειδέα φῦλ' ἀμενηνά.
ARISTOPHANES: *Aves*, 686.

Blut ist ein ganz besonderer Saft.'
. . . .
Die Geisterwelt ist nicht verschlossen;
Dein Sinn ist zu, dein Herz ist todt !
Auf ! bade, Schüler, unverdrossen
Die ird'sche Brust im Morgenroth.
GOETHE: *Faust*.

PLATO in his Phædon represents Socrates as saying in the last hour of his life to his inconsolable followers, "You may bury me if you can catch me." He then added with a smile, and an intonation of unfathomable thought and tenderness, "Do not call this poor body Socrates. When I have drunk the poison, I shall leave you, and go to the joys of the blessed. I would not have you sorrow at my hard lot, or say at the interment, 'Thus we lay out Socrates;' or, 'Thus we follow him to the grave, and bury him.' *Be of good cheer: say that you are burying my body only.*"[2]

Materialism teaches that there is nothing in the universe but matter and its laws; that there is no spiritual substance; and that what is called mind or soul in a man is but a mode of force and motion in matter, and cannot exist in separation from the body.

If materialism is the truth, you and I cannot die as well as Socrates did. If that part of us which thinks and loves and chooses is not separable from our present material frames, our souls are like the electrical charges in the glands of the poor torpedo-fishes, certain to cease to exist as soon as the cells which originate them have been dissolved. On the Peruvian coasts of South America, men drive horses down to the edge of the great deep, in order that they may receive shocks from electric-eels; and sometimes the hoof of a horse will smite the life out of one of his tormentors; and then the wrecked

---
[1] The forty-ninth lecture in the Boston Monday Lectureship, delivered in the Meionaon.
[2] Plato, Phædon, 115; Jowett's Plato, vol. i. pp. 465, 466; Grote's Plato, vol. ii. p. 193.

swimming creature ceases for ever to be an electric battery, because the cells in which the electricity originated are destroyed once for all. Now, materialism is the doctrine that the soul is in some sense secreted by the brain, as electricity is by the cells of the torpedo-fish or electric-eel, and that, when the brain is dissolved, the soul is no more. I do not call this an impious inference, if it be, indeed, an inference fairly deducible from facts; truth is truth, even if it sears our eyeballs; I call it, however, a withering inference. I am not prejudiced against any conclusion reached through clear ideas; but the momentous issues involved in the affirmations of materialism make me anxious to look into these cells, which Häckel and Büchner and Moleschott say originate the soul. Cabanis, as Carlyle narrates with grimmest humour, thought the brain secreted soul as the liver does bile. This philosophy, and the gospel according to Jean-Jacques, were, we know, two of the broadest and blackest of the far-flapping Gehenna wings that fanned the furnaces of the French Revolution.

It is not commonly known, except among specialists in microscopical physiology, that the latest science has something to say to us of immense import as to the relations of matter and life. That theme comes home to the business and bosoms of all men; and, whatever be the verdict of full investigation, all will be eager to face it, who seek, as we do here, whatever is new and true and strategic in religious thought. On the doctrine of organic cells and living tissues, there is surely no book over fifteen years old that is not largely worthless. A text-book on geology, it is often said, is out of date as soon as it is printed. So swift has been the advance of microscopic investigation, that our cell-theory, which began to be elaborated in 1838, has made its supreme advances since 1860. "All life from a cell:" we have heard that doctrine since 1840. "All life from bioplasm," which is the core of the organic cell, we have heard as a scientific truth since about 1860. The first physiological microscopist in the English-speaking world is now Professor Lionel Beale of King's College, London; and his work on "Protoplasm, or Matter and Life," published with elaborate original plates, some of which are of as late a date as 1874, is one of the most important contributions made to knowledge recently by any original investigator of this central question of questions,—whether, when the cells of the brain are dissolved, the soul, like so much electricity developed through them, is dissipated for ever.

You remember, gentlemen, that in Dresden the great picture of the Madonna di San Sisto has an interior which everywhere suggests an ineffable exterior. Many look upon that painting, and study the hushed, shoreless awe and self-surrender of the eyes of the cherubs in the lower part of the transfigured canvas, and do not ask on what the cherubs are looking. But to cause the observer to ask that, is the chief object of this inspired part of the painting. The Madonna di San Sisto was made for an altar-piece. It was intended to stand before burning incense. In a great cathedral its place would be behind the altar, on which incense is burned to ascend to an unseen

THE MICROSCOPE AND MATERIALISM.   37

but near Holy of Holies. It is on the central Ineffable Presence before the picture, and to which the incense rises, that these supernaturally intense eyes of the cherubs are looking. Santa Barbara, as you will observe, divides her adoration between the Son in the arms of the mother and the Unspeakable Unseen before him. Another kneeling figure looks toward what is within, but points to what is without. Even the eyes of the Son and the mother gather mysterious, measureless strength from the Unseen Ineffable to which the incense rises. To me, for one, that which is exterior in this most celebrated painting of all time is more impressive than that which is interior. If you look on the interior, there in the background, and not noticeable at first, but filling all the ambient air behind the mother and the Son, is a cloud made up of innumerable blissful faces of supernatural beings in eternal youth. But when at Dresden, day after day for a month, I studied the painting, I always forgot these in the Central Presence to which the incense ascends; and I went away always in a kind of trance. I know nothing in art that moves me as much as the Unseen Holy suggested before that picture.

Will you follow me long enough to-day, my friends, to find out that this Madonna di San Sisto of Raphael, whose interior suggests an ineffable exterior, is a true analogue of the cell,—God and the soul without, inert matter within,—every movement of the latter pointing to the former as its only adequate cause. Come near enough to this Madonna painting of Almighty God, and you will be convinced that it was the purpose of the Artist to make the interior suggest the ineffable exterior.

When we study living matter with the highest powers of the microscope, and under the lead of the best original investigators, what does the latest science see?

1. That nothing that lives is alive in every part.
2. That the substance of every living organism consists of three parts,
    (1.) Nutrient matter, or pabulum.
    (2.) Germinal matter, or bioplasm.
    (3.) Formed matter, or tissue, secretion and deposit.

As you stand on some murmurous shore of a tropical sea, and pick up a beautifully coloured shell, with its occupant yet in it, you easily perceive a difference between the living and the not-living part of that organism. No doubt the shell grows; and yet, even while the animal bears it about upon his back, parts of the shell are as truly inanimate as they are when afterward the painted wonder lies on the shelf of your cabinet. The shell grows, but not in every part, if it be of mature size. It increases its bulk chiefly by additions of matter at its edges and on its interior; and these increments are made by a process of growth in the softer parts of the organism. We ourselves do not carry very large shells about upon our persons; but the finger-tips are incased in delicate shells, of which by no means every particle is living. It once has been living; but when you pare matter away from the back of a shell, or from the edge of the finger-nail, you find a very great distinction between it and the quick flesh that is touched in a nerve. Four-fifths of the bulk of most organisms, animal and vegetable, is made up of formed matter. Only one-fifth is really alive.

Into the centre of every organic cell there flows a current of nutrient matter, or pabulum; and this may be wholly inorganic. It may be gas; it may be a mineral compound; it may be formed material from meats and fruits. In a cell [referring to a figure the speaker drew upon the blackboard] this nutrient matter is first transformed into living matter, and next the living matter is thrown off as formed material, to make the cell-wall. There are two currents in an organic cell,—one flowing inward, and conveying nutrient matter with it; the other outward, and bearing with it formed material.

In the centre of the cell, by a process that cannot be explained by chemistry or any physical science, the nutrient matter is changed into living matter.

At the outer edge of the cell, formed material accumulates, and is in some cases tissue, in some secretion, in some an osseous deposit.

You have now, I hope, gentlemen, a distinct idea of the three kinds of matter which are to be found in all living organisms,—pabulum or nutrient matter, bioplasm or germinal matter, tissue or formed matter. There are no living organisms, vegetable or animal, that are not made up wholly of these three kinds of matter.

It is only within a comparatively few years that we have been able to demonstrate under the microscope the existence of this distinction between the inner portions of the cell and the cell-wall. Why, Professor Huxley himself, down to 1853, considered the core of the cell as of little importance, and as having no peculiar office.[1] He has changed his opinion now on that point, as on several others concerning the cell-theory; and this fact is not to his discredit at all, because the microscopical study of living matter is advancing so rapidly, that theories of 1850 and 1860 must often be abandoned.

Professor Lionel Beale, who is an accepted authority as to this class of facts, however much his inferences, which I do not now present to you, may be objectionable to materialists, has made large use of a most important process of staining living tissue by a solution of carmine in ammonia. That particular solution makes red whatever is living in a tissue, and does not colour formed material. When you drench a tissue in that solution of carmine in ammonia, you take it out with all the bioplasts stained red. This discovery has been a source of great advances in our knowledge of living tissues, so many of the ultimate parts of which are colourless, and as difficult as water to dissect optically. Fastening the highest magnifying power upon tissue prepared by this carmine process, what do we see?

3. That germinal points, or bioplasts, are scattered so pervadingly through all organic structures that in no organism is there a space one five-hundredth of an inch square without a germinal point, or bioplast.

---

[1] "The Cell-Theory," Medico-Chirurgical Review, October 1853.

THE MICROSCOPE AND MATERIALISM.

We are sure to find, in any piece of living matter of that size, a bioplast that will colour red in a solution of carmine in ammonia.

4. That the germinal points, or bioplasts, are the only living matter.

5. That all formed matter has once been living matter, and so differs totally from inorganic matter.

Every particle of your oyster-shell has once been living, growing matter, although it now is dead ; and yet, although inanimate, it is not inorganic. The shaggiest back of an oyster is matter of a totally different kind from that of the sand and clay and pebbles of which it makes a couch. Every particle of your muscle, nerve, or bone, has once been a bioplast.

I use the word "bioplasm" instead of "protoplasm," because it is a more definite term. It means always that germinal substance which has the power of transmuting not-living into living matter, and of movement, of self-multiplication, and of producing formed material. "Protoplasm" is a word that has been applied to so many different styles of matter, that its indefiniteness in present usage is a frequent source of confusion of thought in biological discussions. "Bioplasm" and "bioplasts" are words which agree well with "biology," the accepted name of one of the greatest of the sciences.

6. That in the cell of an organic tissue the central portion is always a bioplast.

7. That nutrient matter for the bioplasts may consist of inorganic matter, or of formed matter.

8. That the bioplasts convert the nutrient into living matter, and the living into formed matter.

9. That the transmutation of the not-living into the living occurs in the bioplasts instantaneously.

You will read in the older physiologies that all tissues are made up of cells ; and that is, of course, true ; but you must not suppose that it is the latest doctrine that the cell is the object of supreme interest in living tissue. The cell-wall is formed matter. The bioplast is the unit of growth. Bioplasm may exist without an enveloping wall. It may be a bioplast, and not a cell. You may have expected me to say much about cells and the cellular theory; and I am talking about bioplasts and the bioplasmic theory. The theory of bioplasts has superseded the theory of cells, or rather has given to the latter more definiteness ; so that now we speak of cells with meanings derived from bioplasts.

10. That the cell-wall is formed matter, and not alive, and not necessary to the work of transmutation affected by the bioplast.

11. That bioplasts always arise from previous bioplasts.

12. That they have the power of self-movement in any direction.

13. That they are capable of self-subdivision.

14. That each portion of a self-divided bioplast has the same powers as its parent bioplast.

15. That, when dead, bioplasts cannot be resuscitated.

Let us pause here for a moment to notice leisurely the confusion of thought of those who compare this transmutation of the not-living into the living, with the formation of a crystal. I can form a crystal and dissolve it, and form a crystal again out of the solution. I can take two gases, and mix them, and produce water; and then, by an easy chemical process, I can change the water into these two gases; and I can do this, back and forth, any number of times. But, gentlemen, if a bioplast is once dead, it cannot be resuscitated. Materialists talk about the process of life being a kind of "vital crystallisation," whatever that may mean. Be sure that you hold to clear ideas. Revere the orthodoxy of straightforwardness. I want no philosophy, no platform, no pulpit, no dying pillow, that does not rest on rendered reasons. Owen, who fifteen years ago wrote his great work on the "Anatomy of the Vertebrates," opposed in it Darwinism. He called that system as a whole a "guess endeavour." As others were guessing, he himself ventured to guess how the chasm between the not-living and the living might be bridged. Fifteen years ago, Dr. Lionel Beale did not stand as a lion in the way of such guessing. Owen put forward as a possible hypothesis that we shall find out some day that there is "molecular machinery" that accounts for the phenomena of life. He thinks life in its simplest forms may perhaps be compared to the power a magnet exerts when it attracts certain particles to itself, and rejects others. It seems to have the power of selection. You might say that the magnet is feeding itself to see how it draws up to itself metallic dust. But the reply to all that is, You may magnetise and demagnetise your poor iron any number of times; but kill once the smallest living organism, and there is no remagnetising that. You may change your magnet from state to state, as you may change water to gases, and gases to water. *You may braid and unbraid the threads of any inorganic whip-lash again and again, but once unbraid any living strands, and there is no braiding them together again for ever.*

16. That what the bioplasts effect in the transmutation of nutrient into living matter, and of the latter into formed material, chemistry can neither imitate nor explain.

You must not allow yourself to fall into doubt as to the attitude of materialistic philosophers on this proposition. Who is Häckel? He is a materialist. What is a materialist? One who denies that there is any spiritual substance in the universe, and affirms that matter is the only thing that exists. Can Häckel believe in the immortality of the soul? It is a mild statement to say that he must be in grave doubt about it. Can Häckel believe in God? He says in so many words that "there is no God but necessity." What does Häckel affirm concerning the ability of chemistry to bridge the colossal chasm between the living and the not-living? That it is powerless to do so. That it is impotent to explain how inorganic is transmuted into organic matter. There is nothing in chemistry that can produce life. I asked a friend who lately took his degree in chemistry at Gottingen what was thought there about the possibility of producing in the laboratory any parallels to the action of the bioplasts. "We have given up," said he, "the idea that we can make things grow." "Most naturalists of our time," says Häckel, "are inclined to give up the attempt to account for the origin of life by natural causes."[1] Du Bois Reymond says, "It is futile to attempt by chemistry to bridge the chasm between the living and the not-living."

In the bioplast occurs a change which is a sealed volume to the deepest physical science. Here is the not-living, and there is the living; and instantaneously the change of the former into the latter is effected. You look with your microscope upon the centre of the bioplast, and what do you see? Little germinal points arising in the centre, and enlarging. The bioplast seems to boil bioplasts from its centre. It moves. It divides itself here before our eyes [illustrating on the blackboard]. It throbs. You watch it under your microscope. The viscid mass is throwing out a promontory here and a promontory there, against gravitation, and contrary to all we know of chemical force. Suddenly there come great inlets here and there; and soon your one bioplast has made of itself two bioplasts. Each of the new bioplasts continues to receive nutriment; and in its interior the mysterious transmutation of the not-living into the living, and the preparation of formed material, go on again. Each will divide again; and thus, little by little, we find formed matter woven at the edge of these creeping bioplasts into—what? Nerve, bone, muscle, artery. We find the not-living changed into the living, and formed material thrown off—how? So as to produce all the tissues of the body.

Your microscope demonstrates that the little bioplast has not only the throbbing movement, and power of self-multiplication, but of rectilinear movement also. Once this bioplast was here. It threw off formed material; and that formed material flows away behind it as your thread flows from your spindle. It flows away here—as what? As an incipient nerve. But here another group of bioplasts spin, and a thread flows away—as what? As muscular fibre. There you weave your nerve, there your muscle, there your bone, and there your artery. The bioplasts move on; they convert constantly the nutrient material into living matter, and throw off formed material; and when at last this thread is wound, it has a contractile quality. When that is wound, it has the power of transmitting what we call the nervous force; or, when the other is wound, it is the beginning of a bone: when this other, that is the commencement of an artery; or when this other, that is an incipient vein.

We stand in awe before this action of the bioplasts as incontrovertibly indicating intelligence somewhere. If you please, when the egg begins to quicken, must not the whole plan of your eagle, or of your lion, be kept in view from the first stroke of the shuttles? It is something to weave a nerve, is it not? It is enough to keep us on our knees to know that this little mass of colourless, viscid, and, under the microscope, apparently structureless matter, can weave osseous, muscular, and nervous fibres. But what if they cannot only spin these different threads, but also weave them into warp and woof? I am putting before you facts that are not controverted at all. Dr. Carpenter adopts these views in the latest edition of his

---

[1] History of Creation, vol. i. p. 327.

famous "Physiology." They are wholly authoritative statements of what goes on in every living tissue. Among materialists, and anti-materialists, as they walk over this high table-land of science, there is, I assure you, my friends, unanimity as to essential facts at present; and by and by, perhaps, there will be unanimity as to inferences from facts. My belief is, that these facts should be put before all scholars, and not kept from the masses. The members of the legal, clerical, and literary professions, are trained in the logical method as thoroughly as physicists are, and have a right to test reasoning, even where they cannot for themselves verify facts. When I stand here before lawyers, and before learned ministers, and before scholars better informed than I have had opportunity to be on these great themes, I feel, that, although not men of science, you have the right to test the reasoning of science. I am bringing to you here only what are conceded to be facts; and you are competent to test the logic of the facts. It is the right of every mind to look into the logic of whatever touches immortality, the soul, and all that is highest in human endeavour.

It is beyond contradiction that we know that these little points of structureless matter spin the threads, and weave the warp and woof, of organisms. But the bioplasts are of apparently just the same matter in the eagle and in the lion. You look into the centre of the egg of the eagle, and you will see a little mass of colourless, viscid substance, wholly structureless, so far as the highest power of the microscope can reveal its nature. But, when the egg begins to quicken, there is a different segmentation for each of the four great classes of animal forms. All eggs of the class of vertebrates, for instance, begin their development in the same way, and run on in the same way for a while; but your radiates begin another way, and your articulates another. Examined by all the physical tests known to science, bioplasm is the same, however, in your radiate and articulate, and vertebrate.

Take the twittering swallows under the brown eaves, or your eagle on the cliff, or your lion in his lair: the egg, in each case, is the source of life; and, when the quickening begins, there is nothing to be seen at the centre of the egg but this structureless, colourless, viscid bioplasm. Nevertheless, it divides and subdivides, and weaves, in the one case a lion, and in the other a swallow, and in the other an eagle; and I affirm, in the name of all reason, that, from the very first, the plan of the whole organism must be in view somewhere. You know that when a temple is built, the plan of it is in the cornerstone. You know that when the weaver strikes his shuttle for the first time in the finest product of his art, the whole plan of the figures of the web is before him. We see here the bioplasts weaving their threads; we then see them co-ordinating threads and co-ordinating them *so* as, in the one case, to make your swallow, in another case to make your eagle, in another case to make your lion, and in another case to make your man; and why shall we not say, following the law, that every change must have an adequate cause, that somewhere and somehow there is here what all this mechanism needs,—FORECAST?

THE MICROSCOPE AND MATERIALISM.

What are men talking about when they attribute all this to merely "molecular machinery"? Gentlemen, it is out of date to say that "molecular arrangement" accounts for nerve and bone and tissue and artery and vein. It is getting too late to say that merely molecular arrangement accounts for the weaving of organic threads and the interweaving of thread with thread. Will you consider what a complicated process is required to produce that hand of yours, or this eye, or this ear? No doubt strange powers come into existence with the bioplast. Every bioplast is derived from a bioplast : there is your structureless machine, there a little glue-like, colourless matter ; and that is all there is. All life begins in the bioplast ; and every bioplast known to man has been derived from a preceding bioplast. *Out of what, then, came the first one?*

Professor Huxley writes for "The Encyclopædia Britannica" an elaborate article on biology ; and in the opening page of it he says, " The chasm between the not-living and the living the present state of knowledge cannot bridge." Bring materialism to the edge of that chasm. Häckel calls the bioplasts plastids, but confesses that they are mysteries. You find in them complicated processes going forward in apparently structureless matter. You see chemical law apparently set at defiance. The action of material forces appears to be reversed. Häckel, over and over, admits that we cannot produce life, and that we know of nothing but bioplasm that ever has produced it ; but somewhere and somehow in the turmoil of a cooling planet, he thinks, forsooth, that there must have been a cell originated by fortuitous concourse of atoms, or spontaneous generation.

Precisely there is the rock, gentlemen, on which both materialism and the radical form of the evolution theory wreck themselves. There is, I willingly admit, a use, as well as an abuse, of the theory of evolution. Perhaps Häckel and Huxley illustrate its abuse : Dana illustrates its use. But when I stand at the side of the chasm between the not-living and the living, I, for one,—face to face with facts, and all theory put aside,—feel as I felt at Dresden before that Ineffable Holy. I am in the presence of Almighty God. Every change must have an adequate cause ; and the organic living cell must have outside of it a God, and inside of it an immaterial principle, to be accounted for under the law of causation.

Huxley, more cautious than Häckel, says that life is the cause of organisation, and not organisation the cause of life. He has printed that opinion over and over,[1] and never taken it back. Well, if life is the cause of organisation, probably it is safe to say the cause must exist before the effect. At least, that is Nature's logic. *But, if life may exist before organisation, why not after it? I affirm that the microscope begins to have visions of man's immortality.*

Some force forms the parts of an embryo.
That which forms the parts is the cause of the form of the parts.
The cause must exist before the effect.

---

[1] Huxley, Introduction to the Classification of Animals.

The force which forms the parts of an embryo, or of any living organism, exists, therefore, before the parts.

Life is thus the cause of organisation, and not organisation the cause of life.

Life, therefore, exists before organisation.

If it exists before, it may after.

Summarising, then, the latest science analytically, we see in living matter,—

17. That the bioplasts are a colourless, viscid, and apparently structureless substance, and the same in all animals.

18. That they throw off the formed material, *so* that it constitutes nerve, brain, muscle, artery, vein, bone, and all the mechanism of the organism.

19. That, although of the same chemical composition in the eggs of the different animals, they weave tissues such as to produce the different plans of these animals.

20. That their action involves, therefore, both the formation of tissues and their growth according to the needs of the animal.

21. That it involves the production of all those structures, which, in animal and vegetable organisms, exhibit an adaptation of means to ends.

22. That it involves the co-ordination of tissues, secretions, and deposits in the organism.

23. That the plan of the whole organism is necessarily taken into view from the first stroke of the shuttles of the bioplasts that weave it.

Tennyson sings with an emphasis of far-reaching thought :—

" Flower in the crannied wall,
I pluck you out of the crannies ;
Hold you here in my hand,
Little flower, root and all.
And if I could understand
What you are, roots and all, and all in all,
I should know what God and man is."

So we may say in the light of established science :—

Cells in the crannied flesh,
I pluck you out of your crannies ;
Hold you here in my hand,
Little cells, throbs and all.
And if I could understand
What you are, throbs and all, and all in all,
I should know what God and man is.

V.

LOTZE, BEALE, AND HUXLEY ON LIVING TISSUES.[1]

"This seems to me to be as sure a teaching of science as the law of gravitation, that life proceeds from life, and nothing but life."—Sir William Thomson, "Inaugural Address before the British Association," "Nature," vol. iv. p. 269.

The scientific mind can find no repose in the mere registration of sequences in nature. The further question obtrudes itself with resistless might, Whence came the sequences?"—Professor Tyndall "Fragments of Science," p. 64.

PRELUDE ON CURRENT EVENTS.

Our people are about entering on a presidential election in presence of all the other nations who are our guests. *If a man's head, character, and career are each a truncated cone, lacking all the upper zones, he is no fit centennial candidate.* This autumn's choice may be a rudder of the cause of Civil Service reform in many a century to come. Both political parties assert that a great evil exists in the management of our party political patronage; and both call loudly for reform. Is it not the duty of thoughtful men in all the professions, to see to it that gilded demagogism does not teach the people a lie in the smooth name of democracy? We are told that we must beware of an aristocracy of office-holders. We are assured that Civil Service reform, such as both parties demand, may end in the creation of an office-holding class. Which is the worse, to have the great mass of the minor offices in politics the gift of the higher offices, the upper and lower playing into each other's hands, like gift-enterprises and their patrons, or to have the rule established which Washington and Jefferson and Adams and Madison endorsed, that men shall neither be appointed nor removed on the principle that to political victors belong all political spoils, but shall be put into office for ability and availability, and kept there for good behaviour? *Let us take patronage from party, and give it to the people. Vast gift-enterprises in politics are the subtlest threat in the American future.* They call for attention from all scholars, although, perhaps, not for much discussion in the pulpit as yet. Ministers know much of which they do not speak in public. But, in our circles of influence, it is assuredly in our power to turn public thought upon this enormous mischief in the current political life of a yet young nation. Our Woolseys, our Danas, our Tildens, and our Hayeses are united; and shall educated men of all classes not unite the parlour, the platform, and the pulpit on this now strategic theme? On Civil

[1] The fiftieth lecture in the Boston Monday Lectureship, delivered in Park Street Church.

Service reform, or any other great cause, give me a union of the parlour, the pulpit, and the platform, and I will insure a right attitude of the press; and give me a union of the parlour, the pulpit, the platform, and the press, and a right attitude of politics and of the police will follow.

THE LECTURE.

At certain seasons, it was the custom of the Doges of Venice to symbolise the marriage of their city to the sea by casting a ring into the waves. Transfigured marble, Venice stood at the head of the Adriatic, and made the howling, waste, immeasurable brine her servant. But her conquest was one of love, and of the natural superiority of the loftiest spiritual purposes. The sea murmured through her streets : she made it float her traffic. The Mediterranean flashed far and wide; and far and wide Venice made it carry her thought, her enterprise, her beneficence. The modern Venice is religious science: the modern Mediterranean is physical science. Transfigured marble, the loftiest spiritual purposes on earth—wherever they exist—are the city. Far-flashing, immeasurable sea, a waste plain unless ridden by fleets of holy wills and beneficent enterprises—this is physical science. That city purposes to cover that sea with such fleets. The sea and the city rejoice equally in their nuptials. On this occasion I wish, after the manner of the Doges of Venice, to cast into that sea as a marriage-symbol the ring of the living cell.

You will allow me to be elementary; for we cannot approach the mysteries of the microscope with clearness of thought, without attention to some very humble details. Let me ask every gentleman here to look to-morrow morning at the unsharpened edge of his razor in order to form a distinct idea of what the one-thousandth part of an inch is. I suppose a thousand dull razor-edges put side by side might make an inch. Now, under our better present microscopes, how much breadth may such a razor's edge be made to appear to have? We can magnify the one-thousandth part of an inch to the breadth of three fingers, or, exactly speaking, to the length of that line [referring to coloured diagrams exhibited on the platform]. The one-thousandth part of an inch, or the dull edge of your razor magnified twenty-eight hundred times linear, is as thick as your three fingers.[1] When you have a dot only the one four-thousandth part of an inch in diameter, that is, a dot so small that four like it could lie abreast of each other on your razor's edge, and when you magnify that dot four thousand times, it is of precisely the size of this dot, or as large as an English shilling. We are going into a labyrinth, my friends; and I wish you to know what opportunities for exact observation the latest science furnishes. You will hear the assertion, that, under the highest powers of the microscope, protoplasm or bioplasm is apparently structureless. I beg you to look at your razor's edge in order that when you examine bioplasm with a power that magni-

---

[1] Beale's Microscope.

fies twenty-eight hundred times in a linear direction, and know that a line the thousandth part of an inch thick, under that power would be three fingers broad, you may be tolerably certain, that, if there is any structure in the bioplasm that carmine can stain, you will see it. If you are told that this transparent, colourless, and apparently structureless substance is molecular machinery, and that it has purely physical arrangements, which not only weave bone, muscle, artery, vein, and nerve, but can co-ordinate tissue with tissue, and produce wholly by machinery a plant or animal, you must remember that under your microscope, which gives your razor's edge the breadth of your three fingers, all bioplasm appears to be absolutely structureless.

Ariadne, you know, had a clew, a little thread, which she received from Vulcan, and which she gave to Theseus, by the aid of which he safely penetrated the famous labyrinth of Minotaurus. Cultivated men are now thoughtfully walking into a labyrinth far more complicated than that. Philosophy, not for the first time, but with better weapons than ever before, is entering the border-land between the physical and the spiritual, a labyrinth on the border-ground of the two kingdoms of mind and matter; a border on which will be fought the Waterloos of philosophy for a hundred years to come; a border which will be contested as the Rhine never was; a border where soul and matter, God and man, meet; a border where the questions of immortality, of freedom of the will, of moral responsibility, and even of the Divine Existence itself, will be discussed by the iron lips of the best intellectual artillery on the globe. Now we have in this labyrinth an Ariadne clew, and what is it? Why, simply the axiomatic truth, *that every change must have a sufficient cause.* Until the Seven Stars set in the East, men will not give up their belief, that, whenever a change occurs, there must be an adequate cause for it. We are to behold changes occurring in matter, that, under the best microscope, is apparently structureless. We are to behold harmoniously concurrent changes occurring, that when taken together amount to the building up of your hand and nerves and veins, and heart and ear and eye and brain; and not only to that, but to the co-ordinating and adjusting the wants of each one of these to the wants of each of the others. Εκαστω συμμακοι παντες, as the Greeks used to say (all the allies of each): this is the most wonderful fact in the arrangements of the parts of any living organism. Not only the formation of each part, but the co-ordination of part with part in organic structures, is to be explained, without violence to self-evident truth. *We stand before structureless bioplasm, and see it weaving organisms; and we are to adhere, in spite of all theories, to the Ariadne clew, that every cause is to be interpreted by its effects, and that all changes must have adequate causes.*

Before I come to the discussion of the process of carmine staining of living tissues, it is important that I should sketch briefly the history of the cell-theory in physiology.

What right have I to know anything about physiological and microscopical research? How should a minister, who, if born to his calling, is, as many think, neither man nor woman, but something between the one and the other, dare to know anything about the microscope? I notice that the "New York Nation"—a journal which I respect for its culture, but which occasionally takes a merely library view of human affairs—says that it looked over the catalogues of our theological seminaries lately, and did not find, forsooth, that anything important is known in these professional schools about the recent progress of philosophy or physiology. It found by an attentive examination of printed documents,—about as good evidence concerning the theological instruction in our seminaries as tombstones in cemeteries are concerning the characters of those who lie beneath them,—it discovered, after an exhaustive and astute examination of catalogues, that ministers have no acquaintance whatever with philosophy in its latest forms. It did not ascertain that at Princeton Theological Seminary—that mossy, mediæval school—there is a professorship of the relations between religious and other science. At Andover—a little less mossy, possibly, as you think, but yet sufficiently mediæval—there is a lectureship on that subject; and at some near date there may be established there too, God willing, a professorship on that very theme. Unless a man is equipped in what little of logic and metaphysics a Sir William Hamilton and a John Stuart Mill can teach him, he is not adequately prepared for the Aristotelian lecture-room of Professor Park. What shall we say of the thousand sides of the culture of such a man as Schleiermacher, or Julius Müller?

Go to Germany; and what name at this instant leads the philosophy of the most learned land on the globe? What philosopher is read with the most enthusiasm by students of religious and philosophical science in Germany and England and Scotland? Hermann Lotze. Who is he? I am acutely sorry that you have heard of Herbert Spencer, whose star touches the Western pines, and know nothing of Hermann Lotze, whose star is in the ascendant. The most renowned of the modern German philosophers, he is a great physiologist, as well as a great metaphysician.[1] He is the one that is teaching all Germany—he taught me, among others—to look at this border-land with all the reverence with which we bow down before Almighty God. Who is Hermann'Lotze? A man recognised everywhere as thoroughly acquainted with physiology, as Herbert Spencer is not, especially with the latest research. A scholar enriched by the massive spoils of all the German metaphysical systems, and made opulent by all physiological knowledge, and building up with these two sides the colossal arch of a new system, with many a Christian truth at its summit. Although Hermann Lotze, as professor in the philosophical faculty at Göttingen, and one of the higher advisers of the court of Hanover, does not put himself forward as an apologist for any one particular school of religious opinion, he is everywhere regarded as a supporter of that form of Christian philosophy which is now absorbing all established science.

---

[1] See art. on "Hermann Lotze" in Mind, July Number, 1876.

He is a theist of the most pronounced kind. As to evolution, his positions are nearly those of Dana. He is full of scorn for the idea that the Power that put into us personality does not itself possess personality. Carlyle, toward the end of his famous history of Frederic the Great, says there was one form of scepticism which the all-doubting Frederic could not endure. "Atheism, truly, he never could abide: to him, as to all of us," says Carlyle, "it was flatly inconceivable that intellect, moral emotion, could have been put into *him* by an Entity that had none of its own."[1] This inconceivability is the central proposition of Hermann Lotze's philosophy, the most brilliant, the most audacious, the most abreast of the time, of all the philosophies of the globe. You say I am a reactionary evangelical, and that I stand here endeavouring to hold back the wheels of progress. I find that I have been publicly compared in grave print to one of the persecutors of Galileo; not in so many words, but in thought. The truth is, that, instead of being reactionary, this Boston Lectureship is abreast of the latest German investigation. I am proud to say that I have some acquaintance with Hermann Lotze, and that I regard him, as the rising, as Germany regards Herbert Spencer, as the setting, star in philosophy.

Now, gentlemen, to be brief, the cell-theory and its history may be summarised in twelve propositions:

1. In 1838 the microscope was sufficiently perfected to furnish a solid basis for the observation of facts.

2. Schleiden founded the cell-theory, but restricted it to plants. With him the cell consisted of a vesicle and semi-fluid contents.

3. Schwann added to Schleiden's two elements a third,—the nucleus.

Why am I running over this history? Sir William Hamilton never would discuss any great theme without looking back across the record of its discussion in order to obtain the trend of opinion through a long range. Without historical retrospect, we are easily deceived by temporary swirls of opinion. We have yet another clew besides the one of cause and effect: it is the unanimity of experts. A fair statement of the history of the cell-theory will show that the points that are central in the modern form of that theory were established thirty-five years ago, and that there has been unanimity of conclusion as to all the more essential facts.

(1) "This semi-fluid substance," says Schwann, "possesses a capacity to occasion the production of cells."

(2) "When this takes place, the nucleus usually appears to be formed first, and then the cells around it."

You will not fail to remember the distinction between living matter and formed matter, and that nutrient matter is transmuted by the bioplast into living matter, and then thrown off as formed material. But in the cell are nuclei and nucleoli; and the question of questions in the central part of the cell-theory is, whether the bioplasm existed before the nucleus, or the nucleus before the bioplasm.

---

[1] Carlyle, Frederic the Great, book 23, chap. xiv.

Schwann gave as his opinion on that point thirty years ago, that the nucleus appears to be formed by the semi-fluid substance in the cell.

(3.) "*The cell, when once formed, continues to grow by its own individual powers, but is at the same time directed by the influence of the entire organism in such a manner as the design of the whole requires. This is the fundamental phenomenon of all animal and vegetable life.*"

These words of Schwann are more than thirty-five years old, and express the central truth of the bioplasmic theory of to-day.

(4.) "The generation of the cells takes place in a fluid, or structureless substance, which we may call cell-germinating material.[1]

So much for the cellular theory up to 1840.

4. In 1841 Dr. Henle adopted the cell-theory of Schleiden and Schwann, but pointed out the multiplication of cells by division and budding.

5. In the same year Dr. Martin Barry showed the reproduction of cells by division of the parent nucleus.

6. In 1842 and 1846 J. Goodsir confirmed Barry's proposition, and maintained that "the secretion within a primitive cell is always situated between the nucleus and the cell-wall, and would appear to be a product of the nucleus."[2]

7. In 1845 Nägeli showed the comparative unimportance of the cell-wall.

8. In 1851 Alexander Brown proved that the cell-wall is non-essential.

9. In 1857 Leydig first decidedly declared as established science that the cell-wall is non-essential.

10. In 1861 Max Schultze observed that many of the most important kind of cells are destitute of a cell-membrane. He defined the cell as "a little mass of protoplasm inside of which lies a nucleus. The nucleus as well as the protoplasm are products by partition of similar components of another cell." In 1854 Max Schultze had described certain non-nucleated cells, and doubts were thrown on the universality of the nucleus.

11. In 1856 Lord S. G. Osborne discovered the process of the carmine staining of vegetable and animal tissues.

12. By aid of this process Professor Lionel Beale, between 1856 and 1866, so far advanced the knowledge of living tissues, that now his bioplasmic theory at once supplements and supersedes the cellular theory.[3]

Are you shy of accepting the assertion that the cellular theory, of which you have heard so much, has been superseded by the protoplasmic or bioplasmic theory? Here is Häckel himself, who says, "The protoplasm or sarcode theory—that is, that this albuminous material is the original active substratum of all vital

---

[1] Zellenkeimstoff, Schwann, Reports of the Sydenham Society, 1847, p. 39.
[2] Anatom. Memoirs, vol. ii., Trans. of the Royal Soc. of Edinburgh, 1845, p. 417.
[3] James Tyson, The Cell Doctrine; Dr. John Drysdale, Protoplasmic Theory of Life: London, 1874, pp. 12-108.

phenomena—may perhaps be considered one of the greatest achievements of modern biology, and one of the richest in results."[1]

While we abandon to-day the cell-theory in its old form, we retain it in the new form, if we please to put into the doctrine of the cell the idea that the cell-wall is not essential, but that what is essential is the central viscid, transparent bioplasm, or living, germinal matter.

Gentlemen, I am not a bold man, and therefore I have adopted as an inflexible rule, not to trust any man's authority as to facts in science without advice to do so from his determined opponents. It would have been enough for me to have had, as I did have, the authority of James Dana for trust in Professor Lionel Beale's statements of facts concerning living tissues. One of the most distinguished theological scholars in this country, whom, out of reverence, I will not name, was afflicted nervously, and threatened with loss of sight. Physicians in this learned city, and in Paris, again and again prescribed for him, but fruitlessly. Dr. Lionel Beale in London was recommended to him; and one hour of examination of the case was followed by a single prescription, which was effectual, and has been so year after year through a quarter of a century. In one of my groves near Lake George there is a beech which I call "The Bioplast Beech," so delicious were the hours I spent there this summer with Hermann Lotze and Beale and Dr. Carpenter and Dana and Darwin, and a score of other books of science. Beale's celebrated lectures before the Royal College of Physicians in 1861, on living tissues, and his discoveries concerning bioplasm, were preceded by a work on "The Microscope," which you had better not buy yet, simply because it is going into a fifth edition. It is a bulky, elaborate book, full of plates; and I have seen it worn ragged in my library, as I call the Athenæum yonder, with its one hundred thousand volumes, its one hundred magazines, and one hundred newspapers and excellent professional collections. It is a significant sign when a book of science is worn ragged in a library used by the Sumners and Wilsons and Emersons, and other men who are not likely to waste time on rubbish.

Beale's volumes I find worn eloquently black, and Bastian's hardly stained. Some small philosopher may tell you that Beale is no authority, and that many of his propositions are in dispute. One of them is; but it is a proposition that I am not using at all, namely, that the nerves end in loops. Even on that obscure point, opinion is turning more and more to Beale's side. But when a costly work on the miscroscope, with elaborate plates filled with the results of original research upon living tissues, goes in a few years into a fifth edition, and its author is commonly pronounced to be the first microscopist of the English-speaking world, and when its facts agree with those of Frey, the greatest authority on the same subject in the German-speaking world, even a timid man may read such a book without any great tremor. In examining authorities in science, I seek first to ascertain on what points there is an agreement of the

---

[1] Häckel, Quar. Mic. Jour., 1869, p. 223.

best English and the best German publications; but that is not enough. We must have the authority of his rivals for trusting any man as an expert.

What do the opponents of Beale's conclusions say of his facts?

1. Dr. John Drysdale of Edinburgh is the author of a work on "The Protoplasmic Theory of Life;" and in 1874 was president of the Liverpool Microscopical Society. He has given head and heart to the doctrine that bioplasm is a form of matter *sui generis;* and that its activity is an outcome of transmuted physical force, or the result of "irritability under stimulation."

He opposes vehemently Beale's conclusion that the actions of bioplasm require to account for them a higher than physical force. But of Beale he says, "A master-mind appeared in 1860, we are glad to say, in the person of our countryman, Dr. Lionel Beale of London. He had for years devoted himself with unwearied zeal to microscopical research on the animal tissues, using the highest magnifying powers as soon as available, and had attained to an almost unrivalled skill, and had discovered various new methods of the preparing objects, which enabled him to analyse the structures of the textures to a point not hitherto reached by anatomists. In 1860 he wrote those 'Lectures on the Structure of the Simple Tissues of the Human Body,' which were delivered before the Royal College of Physicians in 1861, and which are destined, I believe, to make an epoch in the progress of physiological science. Since then, Dr. Beale has gone on completing and expanding his system, and filling up the details, and has carried it out in pathology to an extent of completeness and consistency marvellous for the short time as yet given, and as being the work of one man; a fact which in itself shows he has seized on one great and central principle, which enables him to bring into practical harmony a vast number of scattered observations both of his own and of others. Beale's protoplasmic theory now takes the place of the cell-theory. *General opinion is now in accord as respects the facts with Dr. Beale's statements on the nucleus in 1860.*"[1]

2. Professor Alexander Bain makes Beale's facts the basis of the central chapter in his work on "Mind and Body,"—one of those tempting but disappointing royal roads to knowledge called "The International Scientific Series." Bain, as you know, teaches that only matter exists in the universe, but that matter rightly defined is "a double-faced somewhat, having a spiritual and a physical side." That is the nearest approach to a definition that either he or Tyndall has given. In this marvellous compound unit there coinhere in one substratum extension and the absence of extension, form and the absence of form, activity and the absence of activity,—all the perfectly contradictory attributes of matter and mind. I suppose that it may be asserted that mind is coextensive with matter; but never, until we can believe that a thing can be and not be at the same time and in the same sense, will men who love clear ideas adopt Tyndall's and Bain's self-contradictory definition of matter.

---

[1] Dr. John Drysdale, Prot. Theor. of Life: London, 1874. Pp. 41, 68, 103.

LIVING TISSUES. 53

But even Bain leans confidently on Beale whenever he speaks of microscopical physiology.

In arguments before juries, Webster often asked his opponents, "Why do you not meet the case?" Remember that famous phrase of his, if you hear the materialistic theory of evolution defended. What is the case against that theory? It consists of the irreconcilable opposition of the attributes of matter and mind, of the unfathomed gulf between the not-living and the living, of the fact that spontaneous generation has never been shown to be a possibility, and of the missing links between men and apes. Let these points be met fairly, and the case is met. Not until the chasm between the not-living and the living is filled up by observation, not until that distant time when you shall have found some merely physical link between the inorganic and organic, can you say that *the* theory of evolution has been proven by induction. A theory of evolution has been proved, but not *the* theory. The public mind is immensely confused by this one word of many meanings. A theory of evolution Dana holds, but not *the* theory. The position of this Lectureship is, that there is a use and an abuse of the theory of evolution, and that Häckel illustrates the abuse, and Dana the use. I hold *a* theory of evolution, but not *the* theory. What do I mean by *the* theory of evolution? Precisely what Huxley means when he says in so many words [1] that "if the theory of evolution is true, the living must have arisen from the not-living."

3. You want Huxley himself in support of Beale, and you shall have him. The most important propositions that I shall present to you on this occasion I hold here in my hands; and they are all in the language, though not in the order of statement, which Professor Huxley uses. I do not know any late leading work in Germany on microscopical physiology that does not mention Beale again and again. When I was in Jena, I bought Ranke's great work on physiology, in spite of the fact that I was a minister who had no right to know anything on this subject. I brought it with me across the Atlantic; and, on opening it the other day, I found Beale cited, and his propositions put into the foreground of the latest German statements of the cell-theory. You know that, Schleiden and Schwann being Germans, the German physiologists, from patriotic and various other motives, cling to the nomenclature of these great men; but they honour Beale. When I turn to Huxley, however, in his article on biology, in the latest edition of the twenty-one volumes of "The Encyclopædia Britannica," I am able to select from various parts of his discussion these seventeen propositions, every one of which was first made sure by the microscopic research of Lionel Beale; but Beale is not once mentioned in this article by Huxley.

1. "It is certain that in the animal, as in the plant, neither cell-wall nor nucleus are essential elements of the cell."

That conclusion is the result of a Waterloo battle, if you please. Although the proposition is so quietly stated, Huxley knows what

---

[1] Encyc. Brit., ninth ed., art. Biology.

proof there is behind it, and lays it down before the world in this, his most scholarly production on biology, and his latest, as established science.

2. "Bodies which are unquestionably the equivalents of cells—true morphological units—are sometimes mere masses of protoplasm, devoid alike of cell, wall, and nucleus."

3. "For the whole living world, then, it results that the morphological unit, the primary and fundamental form of life, is merely an individual mass of protoplasm."

4. "In this no further structure is discernible."

I beg you to notice the accord of all these propositions with those which, in the last lecture, I put before you as the result of Lionel Beale's investigation.

5. "The nucleus, the primordial utricle, the central fluid, and the cell-wall, are no essential constituents of the morphological unit, but represent results of its metamorphosis."

We saw how bioplasm throws off formed material, and how the nucleus is the result of the action of the bioplasm, and not bioplasm the result of the nucleus; and here you find Professor Huxley asserting that the nucleus is a result of the metamorphosis of bioplasm.

6. "Though the nucleus is very constant among animal cells, it is not universally present."

7. "The nucleus rarely undergoes any considerable modification."

8. "The structures characteristic of the tissues are formed at the expense of the more superficial protoplasm of the cells."

The structures characteristic of the tissues! What a smooth phrase that is, for the infinity of design in the human constitution, bone, nerve, artery, muscle, and all that makes a plant a plant, or an animal an animal!

9. "When nucleated cells divide, the division of the nucleus, as a rule, precedes that of the whole cell."

10. "Independent living forms may present but little advance from an individual mass of protoplasm."

11. "All the higher forms of life are aggregates of such morphological units or cells, variously modified."[1]

12. "The protoplasm of the germ may not undergo division and conversion into a cell aggregate, but various parts of its outer and inner substance may be metamorphosed directly into those physically and chemically different materials which constitute the body of the adult."

13. "The germ may undergo division, and be converted into an aggregate of cells, which give rise to the tissues by undergoing a metamorphosis of the same kind as that to which the whole body is subjected in the preceding case."[2]

14. "Sustentative, generative, and correlative functions in the lower forms of life are exerted indifferently, or nearly so, by all parts of the protoplasmic body."

---

[1] Prof. T. H. Huxley, Encyc. Brit., ninth edit., Biology, pp. 681, 682.   [2] Ibid., p. 682.

15. "The like is true of the functions of the body of even the highest organisms, so long as they are in the condition of the nucleated cell."[1]

16. "Generation by fission and gemmation are not confined to the simplest forms of life. Both modes are common, not only among plants, but among animals of considerable complexity."

"*Throughout almost the whole series of living beings, we find agamo genesis, or not-sexual generation.*" "Eggs, in the case of drones among bees, develop without impregnation."[2]

[After a pause, Mr. Cook proceeded in a lower voice],—

When the topic of the origin of the life of our Lord on the earth is approached from the point of view of the microscope, some men, who know not what the Holy of Holies in physical and religious science is, say that we have no example of the origin of life without two parents. There are numberless such examples. "When Castellet," says Alfred Russel Wallace, Darwin's coadjutor, "informed Reaumur that he had reared perfect silk-worms from the eggs laid by a virgin moth, the answer was, '*Ex nihilo nihil fit,*' and the fact was disbelieved. It was contrary to one of the widest and best-established laws of Nature; yet it is now universally admitted to be true, and the supposed law ceases to be universal."[3]

"Among our common honey-bees," says Häckel,[4] "a male individual, a drone, arises out of the eggs of the queen, if the egg has not been fructified; a female, a queen, or working-bee, if the egg has been fructified."

Take up your Mivart, your Lyell, your Owen, and you will read this same important fact which Huxley here asserts, when he says that the law that perfect individuals may be virginally born extends to the higher forms of life. I am in the presence of Almighty God; and yet—when a great soul like the tender spirit of our sainted Lincoln, in his early days, with little knowledge, but with great thoughtfulness, was troubled by this difficulty, and almost thrown into infidelity by not knowing that the law that there must be two parents is not universal—I am willing to allude, even in such a presence as this, to the latest science concerning miraculous conception.

17. "The phenomena which living things present have no parallel in the mineral world."[5]

What now, gentlemen, is the conclusion of Huxley from all these propositions that seem to point one way? You notice that his facts are Beale's. You find an explicit agreement here of Beale, of Huxley, of Bain, of Drysdale, of Ranke, and I might say of Carpenter, of Dalton, and of scores of recent specialists. The facts being established, the supreme question as to their interpretation is,—Life or mechanism, *which ?*

Beale says life: Beale says a principle that cannot be explained

---

[1] Professor T. H. Huxley, Encyc. Brit., ninth edition, Biology, p. 685.
[2] Ibid., pp. 686, 687.
[3] Alfred Russel Wallace, Miracles and Modern Spiritualism, p. 38 : London 1875.
[4] History of Creation, vol. i. p. 197.   [5] Ibid., p. 68

by any form of merely physical force. But Huxley says, and be amazed all men who hold the Ariadne clew, "A mass of living protoplasm is simply a molecular machine of great complexity, the total results of the working of which, or its vital phenomena, depend, on the one hand, on its construction, and, on the other, upon the energy supplied to it : and to speak of 'vitality' as anything but the name of a series of operations is as if one should talk of the horologity of a clock." You are shocked at this proposition, and therefore I have not spoken in vain. We will consider next week this astounding *non sequitur*. If Hermann Lotze, the first philosopher of Germany, were on this platform to-day, he, in the name of the axiom that every change must have a sufficient cause, would tear into shreds the materialistic or mechanical theory of the origin of living tissues and of the soul.

VI.

LIFE, OR MECHANISM—WHICH?[1]

"Tu cuncta superno
Ducis ab exemplo, pulchrum pulcherimus ipse
Mundum mente gerens, similique imagine formans."
BOETHIUS, *De Consol.*, 9.

"What time this world's great workmaister did cast
To make all things such as we now behold,
It seems that He before His eyes had plast
A goodly patterne, to whose perfect mould
He fashioned them as comely as He could,
That now so fair and seemly they appear;
As naught may be amended anywhere.
  That wondrous patterne, wheresoe'er it be,
Whether in Earth, laid up in secret store,
Or else in Heaven, that no man may it see
With sinful eyes, for fear it to deflore,
Is perfect beauty."—SPENSER.

ONE day the poet Goethe, when in his advanced age, was riding home to Weimar with his friend Eckermann, and conversing on the immortality of the soul. They turned by Tiefurt into the Weimar road, and stopped at a spot, where, like other travellers, I have often meditated on Goethe's career; and they had from that outlook a majestic view of the setting sun. The great poet and philosopher remained for many minutes in perfect silence, and at last said with mystic but tremorless emphasis, "*Untergehend sogar ist's immer dieselbige Sonne.* Setting, nevertheless the sun is always the same sun. I am fully convinced that our spirit is a being of a nature quite indestructible, and that its activity continues from eternity to eternity." This man knew all philosophies and all art—materialism, realism, pantheism, the wildest scepticism, and, I fear, not a little of the most infamous sensualism; but his was at least a free mind and a modern one. Here, however, was his conclusion concerning the possibility of the existence of the soul in separation from the body: *Setting, nevertheless the soul is always the same soul.*[2] Will you enter to-day, my friends, into Goethe's brain at that instant, and remain

---

[1] The fifty-first lecture in the Boston Monday Lectureship, delivered in the Park Street Church.
[2] Goethe, Conversations with Eckermann, Trans. by J. Oxenford, Bohn's ed., p. 84.

there during this discussion, lynx-eyed, I care not how thoroughly so, but earnest? It is incontrovertible that we, too, a little while ago, were not in the world, and that we, too, a little while hence, shall be here no longer. The sun hastes to the west as fast at noon as in the last moment before sunset.

New lands in our age can be discovered only in old lands. Schliemann, on the Plain of Troy, has shown us a city of great antiquity; and he has done so by studying an old land beneath its soil. We are reaching the bottom of the Roman forum; we understand, as never before, the environment of the Acropolis, because we are looking with the spade for new lands in the old lands. If a new continent has been discovered anywhere in the last twenty-five years, it has been in the ancient continent of *living tissues*. We are to enter on that strange country; we draw near to it across turbulent seas; and I think, that, as the Santa Maria ploughs tossing across the waves toward the west, we already begin to see carved wood occasionally, symbol of life behind the watery horizon. Already, as we approach this new continent, do we not find now and then a poor floating spray of red berries? Are these little birds not of a kind always cradled on the land? Are not the shapes of the very clouds, as the sun goes down, some indication that we shall at last reach the firm, happy shore? Is there not breathed upon us out of the undescried but nearing coast an odour as of spices and balm, and frankincense and myrrh, and dates and palms—a fragrant atmosphere that comes in the twilight wind off the continent of an unseen Holy? We have not landed on the new coast yet; but they who walk late on the deck of the Santa Maria have seen a light rise and fall ahead of us. We are to look to-day at the thickening signs of the approach of a whole new continent in philosophy that lies hardly out of sight. It will be a land assuredly of firm hope of immortality, and therefore a land of inspiration such as no spiced breath of the tropics ever breathed into the physical nostrils. Our souls are sick from lack of the more heavily fragrant airs out of the blessed isles of certainties as to what is behind the veil. It is already certain that we are to discover a new land, and that the inhabitant of it is life, not mechanism.

Two positions of much importance have been proved, I hope, in lectures preceding this: first, the explicit and entire agreement of Beale and Huxley as to all the central facts concerning living tissues, and this in spite of the disagreement of these authorities on other points; and, secondly, the crescent unanimity of experts for thirty-five years as to those same facts. The two initial propositions which I think I have established are, that rival experts agree, and that they have agreed for more than a quarter of a century, on the facts fundamental in our discussions here. Let us, now, summarise our knowledge of bioplasm, remembering, as we do so, that we have the authority of Huxley, of Carpenter, of Frey, of Dalton, of Beale, of Drysdale, of Bain, of Ranke, and of Kölliker. You will permit me, for the sake of clearness of thought, to number the points of our positive knowledge in biological science.

Bioplasm, otherwise called protoplasm, or germinal matter,—
1. Is transparent ;
2. Colourless ;
3. Viscid, or glue-like ;
4. Under the highest microscopical powers is apparently structureless ;
5. Exhibits these characters at every period of its existence ;
6. Shows itself, under all the tests known to physical science, to be the same in the animal and in the plant, in the sponge and in the brain ;
7. Is capable of throbbing movements, or of advancing one portion of itself beyond another portion ;
8. Is capable of rectilinear movements ;
9. Executes so many movements, that the same mass probably never twice in its life assumes the same form ;
10. May exist in masses less than one one-hundred-thousandth of an inch, or as large as one two-hundredth of an inch in diameter, but, as constituting the nuclei of fully-formed cells, is usually found in masses from one six-thousandth to one three-thousandth of an inch in diameter ;
11. Absorbs nutrient matter, which may be either inorganic or formed material ;
12. Instantaneously changes this dead matter into living matter ;
13. Does so by a process which no human science can imitate or explain ;
14. Throws off formed material to constitute a cell-wall ;
15. Develops within itself a nucleus, and within that a nucleolus ;
16. May exist and move, however, without cell-wall or nucleus ;
17. Spins the threads of nerves, arteries, veins, bones, and all the mechanism of the system, by throwing off formed material ;
18. Weaves these threads into the infinity of co-ordinated designs in the plant and animal ;
19. Can by no possible outer environment be made to produce nerve if it should produce muscle, or muscle if it should produce nerve, and so of every other tissue, secretion, and deposit ;
20. Is so thickly scattered through the tissues, that there is scarcely a space one-five-hundredth of an inch in size without its portion of it ;
21. Is capable of self-subdivision ;
22. In its self-subdivided parts has all its original powers ;
23. Always arises from preceding bioplasm ;
24. Constitutes about one-fifth of the bulk of living bodies ;
25. Is the sole agency by which every kind of living thing is made, or, so far as known, has been made or ever will be made ;
26. When it divides itself, is preceded sometimes in that act by the division of its nucleus, and sometimes not ;
27. May throw off a portion of itself without a nucleus, and develop a nucleus in the detached portion.
28. Forms nuclei and nucleoli, which appear to differ sexually, as it is only after the intermingling of these in certain cases that multiplication takes place ;

29. Does not transform the nucleus, or nucleolus, directly into formed material;

30. Transforms it into ordinary bioplasm, and thus into formed material;

31. When recently dead, will take a carmine stain from the solution of carmine in ammonia, as formed material will not;

32. At its death is resolved into fibrine, albumen, fatty matter, and salts;

33. Forms thus the spontaneously coagulable substance on the diffusion of which through the body the rigidity of the frame after death depends;

34. Is in direct continuity with formed material while the latter is in process of formation.

Such is the most interesting, by far, of all the objects known to physical science.

Carmine staining, the great discovery of 1856 and 1860, must take place immediately after the death of the bioplasm, or it cannot be successfully executed. Many unskilful manipulators in the laboratory, and amateurs without number, have endeavoured to stain the tissue of plants and animals, and have waited too long after its death, and have failed. Sometimes, too, they have not rightly compounded the materials for their carmine solution, a distinct receipt for which you will find in Beale's work on the microscope. When the process of staining is performed soon after the death of a tissue, all germinal points or bioplasts in it come out with a red colour; but the formed material is not stained at all.

[From this point on, Mr. Cook referred to large coloured diagrams hung on the wall back of the platform.]

These eloquent representations of stained tissues are exact reproductions of Dr. Beale's famous illustrations, and were made by Mr. Stone, an artist of the Studio building, who spoke admiringly of Beale's illustrations the instant he saw them. Here is the whole cell with its wall, bioplast, and nucleus (see plate I. fig. 1). Two currents exist in every cell,—one flowing inward in the direction of this arrow, and the other passing out from the centre of the bioplast in the direction of this arrow. Every particle of matter that can be found in a living being is of one of three kinds,—nutrient matter, living matter, or formed matter. Nutrient matter comes through the wall of the cell, and, entering into the bioplasm, is there transformed into living matter.

You had better not take a cell, however, as the type of the elementary part in the living tissue. If you are to be abreast of the very latest investigations concerning the cell-theory, you will take a naked mass of bioplasm like this as the elementary part (see plate I. fig. 2). As I showed you in my last lecture on both Huxley's and Beale's authority, it is not essential at all that there be a wall of formed material around the naked mass of bioplasm. It is not essential at all that there be a nucleus within it. That is the advance we have made since 1838. Nevertheless, if you are to understand the action of these currents, it is well to keep in mind the cell-wall. Nutrient material may pass through the cell-wall in animal tissues just as sap passes through the intercellular substance in vegetable tissues. When once in the bioplast, the nutrient matter is seized on by this living matter, which you see coloured with carmine in all these illustrations, and nuclei are developed in the bioplast, and nucleoli within the nuclei. The bioplast produces the nucleus, and not the nucleus the bioplast. It throws off formed material around its quivering edges, and thus forms a cell-wall. In that wall the oldest formed material is on the outside, and the next oldest just within, and so on to the inner part of the wall, which is in physical continuity with the bioplasm.

Movement is going on all the while in any naked mass of bioplasm. Here is a bioplast, naked, colourless, structureless matter; and it moves so that it takes these many shapes in five seconds, and these many other shapes in one minute (see plate I. figs. 2 and 3). Here we must hold fast to the Ariadne clew, that every change must have an adequate cause. We come here to fathomless design; but let us enter by slow stages on these sublimities of research.

Here is a young tendon, and here is an old tendon. The living matter is red, as you notice, and runs in lines through the tendon; and yet the tendon is narrow. But in the old tendon the formed material is more abundant than in the new; and yet all the formed material which makes an increased thickness in the old has been thrown off by these bioplasts. They have here thrown off formed material *so* as to make a tendon, which is, as you know, a structure very different from muscular fibre and from nervous fibre.

Here is one set of bioplasts that is intended to weave a tendon, here one that is to weave a muscular fibre, and here one that is to weave a nervous fibre. There is no possible external influence that can make them exchange offices with each other. You have here a tendon, there a muscle, there a nerve, all woven by these bioplasts. We know that they are thus woven, and that every change must have an adequate cause. Adhere, gentlemen, to that axiomatic truth, though the heavens fall. From your bioplast spindles flows off formed matter—here a miracle of muscle, there a miracle of tendon, there a miracle of nerve.

The cellular integument is not unworthy of notice; for that shows us the career of its bioplasts from the first to the last. You have here the skin that covers one of the papilla on the tongue of a frog (see plate II. fig. 1). That infinitely delicate membrane that covers the little sensitive points on the tongue is here magnified. You notice that the bioplasts on the lower or inner side are young, and that there is not much formed material around them. There are no distinct cells in the younger part of a tissue. This intercellular substance is not formed into the ring-shapes which you see further on, where the tissue is older. As the bioplasts grow, the formed material about them increases in thickness, until it becomes so thick that the nutrient matter will not go through the cell-walls. Then the bioplasts languish; they grow smaller and smaller, and at last the cells in which the bioplasts are dead scale off. When dead—never before except by violence—they drop away; but their places are supplied by soldiers that take position in the gap of the lines, and build according to the pattern of the design of the whole organization. You have here (see plate II. fig. 2) coloured illustrations of several stages of the growth of a cell—its youth, its adolescence, its middle life, its advancing age, its extreme old age.

Remember that a mass of bioplasm has a tendency to assume a more or less spheroidal form. But it changes itself in the course of a minute into all the protean shapes indicated here, first by the black, then by the unbroken line, then by the broken red line, and divides and subdivides its edges, until at last it throws off this portion of itself, which has the same powers with its parent (see plate I. fig. 3). We find under our astounded gaze nothing but colourless, glue-like, transparent matter; and yet we see it performing all these miracles of as many different sorts as there are different sorts of tissues to be woven.

In a single nerve there is an unspeakable complexity; but come to something a little more complex. Let us stand with open eyes before this revelation of Almighty God. Here is a nerve wound spirally around another fibre (see plate II. fig. 5). How is it made to twine about its trellis-work? Why, when that nerve begins to be formed in a living organism, these bioplasts in it are near each other. They begin to throw off formed material. The object is to weave so as to produce this delicate nerve that is coiled spirally around the other fibre. The bioplasts were shoulder to shoulder, and they begin to separate. They weave, and they carry a spiral nerve around that other fibre with perfect precision.

Adhere to your clear ideas. Materialists say that all this is done by molecular machinery. Do they know what they are talking about when they use that phrase? They say that here are "infinitely complicated chemical properties." They say that all these things occur merely by "a transmutation of physical forces." Do they know what they are saying when they utter propositions of that sort? The tendency of the latest science begins to throw into derision all materialism of this kind. The Germans have a proverb which says, "The clear is the true;" and ascertained truth can be made clear. Will you make it clear that "molecular machinery," however complicated, can achieve these results? There a tendon, there a muscle, and there a nerve, are woven, and all by the same machinery? The same causes ought to produce the same results. There is an almost measureless difference in your results; but in all ascertainable physical qualities this bioplasm is the same thing in every tissue.

Marvels, however, have but just begun. We might pause long on these earlier stages in the formation of tissues; but there is one word or fact we ought to bow down before, if we have eyes (see plate III.). It is *co-ordination*, the adjustment of part to part in a living organism. A vast number of tissues are woven side by side; and their *co-ordination* is the supreme miracle. It is more than much, my friends, to weave a nerve, a muscle, a vein. But here we have a mass of thin tissues from a tree-frog, and you have here muscles and veins and nerves interlacing with each other intricately. Not only do the mystic bioplasts know enough to coil one fibre around another fibre spirally, but they weave the whole complexity of the tissues together. How ? So that there is no clashing among the multitudinous wheels of the living organism. In the naked bioplast we see changes going on ; and the question is, What is an adequate cause of these changes? Life, or mechanism—which? In the different threads that are woven by the bioplasts we must ask : Life, or mechanism—which ? But here, before this transfigured representation of the co-ordination of tissue with tissue, the question answers itself : Life, or mechanism —which ?

Here is the last white and mottled bird that flew to us out of the tall Tribune tower ; and softly folded under its wing are these words concerning Darwin from Thomas Carlyle at his own fireside in London : " So-called literary and scientific classes in England now proudly give themselves to protoplasm, origin of species, and the like, to prove that God did not build the universe. I have known three generations of the Darwins,—grandfather, father, and son, atheists all." [I do not call Darwin an atheist ; but this testimony is very significant.] " The brother of the present famous naturalist, a quiet man, who lives not far from here, told me that among his grandfather's effects he found a seal engraven with this legend, ' *Omnia ex conchis* ' (' everything from a clam-shell '). I saw the naturalist not many months ago ; told him that I had read his ' Origin of the Species,' and other books ; that he had by no means satisfied me that men were descended from monkeys, but had gone far toward persuading me that he and his so-called scientific brethren had brought the present generation of Englishmen very near to monkeys. A good sort of man is this Darwin, and well meaning, but with very little intellect. Ah ! it is a sad and terrible thing to see nigh a whole generation of men and women professing to be cultivated, looking around in a purblind fashion, and finding no God in this universe. I suppose it is a reaction from the reign of cant and hollow pretence, professing to believe what in fact they do not believe. And this is what we have got : all things from frog-spawn ; the gospel of dirt the order of the day. The older I grow,—and I now stand upon the brink of eternity,—the more comes back to me the sentence in the Catechism, which I learned when a child, and the fuller and deeper its meaning becomes,—' What is the great end of man ? To glorify God, and enjoy Him for ever.' No gospel of dirt, teaching that men have descended from frogs through monkeys, can ever set that aside." [1]

Will haughty Boston, will the colleges of New England, will tender and thoughtful souls everywhere, listen to Thomas Carlyle as he stands upon the brink of eternity ?

---

[1] Daily Tribune, Nov. 4, 1876. Extract from a letter from Carlyle published in Scotland, and quoted in the London Times.

VII.

DOES DEATH END ALL? INVOLUTION AND EVOLUTION.[1]

"Die Nothwendigkeit für zwei unvergleichbare Kreise von Erscheinungen zunächst zwei gesonderte Erklärungsgründe zu verlangen, verbot uns jeden Versuch, aus Wirkungen materieller Stoffe, so fern sie materiel sind, das innere Leben als einen selbstverständlichen Erfolg ableiten zu wollen."—HERMANN LOTZE, *Mikrokosmus*, I., 186.

"Attention to those philosophical questions which underlie all Science, is as rare as it is needful."—Professor T. H. HUXLEY, *Contemporary Review*, Nov. 1871, p. 443.

IF the Greeks had possessed the microscope, they would in all probability never have been thrown into debate over the famous question of their philosophy, whether the relation of the soul to the body is that of harmony to a harp, or of a rower to a boat.[2] According to the former of these two theories, the music must cease when the harp is broken: according to the latter, the rower may survive, although his boat is destroyed. He may be completely safe, even when his frail vessel, splintered by all the surges and lightnings, rots on the tusks of the reefs, or sinks in the fathomless waste, or dissolves to be blown about the world by the howling seas. In the one case, death does, in the other it does not, end all. Dim as was to the Greeks of Pericles' day the whole field which science has entered with the microscope for the first time in the last fifty years, all their greatest poets and philosophers held that the relation of the soul to the body is that of the rower to a boat. This was the common metaphor as men conversed on this theme under the Acropolis two thousand years ago. Without Christian prejudices, Greek tragedy is full of the dying faith of Socrates. Æschylus, with his eyes of dew and lightning fixed on the fact of immortality, strikes the central chord of his harp; and one terrific thrum of it I often in still days hear across twenty centuries:

"Blood for blood, and blow for blow:
Thou shalt reap as thou didst sow."

---

[1] The fifty-second lecture in the Boston Monday Lectureship, delivered in Tremont Temple. [2] Plato, Phædon.

DOES DEATH END ALL? 65

What if Aristotle and Plato and Æschylus had had Beale's and Helmholtz's and Dana's eyes in the study of living tissues?

When modern investigation asserts that life directs the movements of bioplasm, it does not deny at all that currents of physical and chemical forces are floating around the bioplast boat. It asserts simply that the oars are in the hands of life. You will not understand me to deny that the rower in the boat is aided by the currents beneath him, by the winds around him, and by his own weight and the inertia of his vessel. Nevertheless, between the rower and the boat on the one hand, and the inert log that may be floating beside him on the other, there is plainly all the difference that exists between the living and the not-living. Your rower takes advantage of all the forces around him; he can give them new directions; he presides over them. He can sail against the wind; he can row against the current; he governs the forces that wheel in mysterious complex cycles above and around and beneath him; he makes them his own, and so is a living thing on the water. Just so, life uses the physical and chemical forces at work in living organisms.

There ought to stand before every discussion definitions, just as before one of Shakspeare's dramas there stand the names of the *dramatis personæ*. I know into what an intricate tropical forest of thought I am entering; and I am fully aware that the chief personage here is one whose character never has been successfully described in a definition. What is life? Thousands and thousands of definitions have been attempted of that term; and we have as yet in words no satisfactory statement of what life means; but we all understand very well what the thing is.

Herbert Spencer defines life as "The definite combination of heterogeneous changes, both simultaneous and successive, in correspondence with external co-existences and sequences." This definition has been very much admired; and I suppose you all understand what it means. The latest science finds this definition *defective*, because it does not limit the changes of which it speaks to one specifically constituted substance now known as bioplasm.[1]

I know what I venture; but, as my definition of life, I must give these words: *The power which directs the movements of bioplasm.* I beg you to notice that I do not say that life is *the force which moves bioplasm*, although, as a loose definition, the latter phrase would do. Bioplasm is moved in part by physical and chemical forces, though not chiefly. Chemical and physical forces, however, are not called living in the best philosophy. To say that life is the force that moves bioplasm is to say that all the power there is in the river on which the boat and rower float originates in the rower. I say nothing of that sort. The force of the river belongs to the river; that of the oars, to the rower. The power which causes your skiff to move against the current, or which catches the wind in the sail, is that of its living occupant, who directs other forces, *and puts forth force of his own*. Nevertheless, in the motion of your little boat, there is a

---

[1] Drysdale, Protoplasmic Theory of Life: London, 1874. P. 176

combination of the power of the rower and the power of the currents. So, in the motion of your bioplast, there is the agency of purely physical and chemical forces, together with the co-ordinating agency or directing power which weaves the tissues, and interweaves tissue with tissue into designs marvellous beyond comment, and which cannot be accounted for at all by anything simply chemical or physical. I affirm, therefore, that life may be defined provisionally as the rower in the boat, or the power which directs the movements of germinal matter. To give a fuller definition, I may say that *life is the invisible, individual, co-ordinating cause directing the forces involved in the production and activity of any organism possessing individuality.* Of course the vitality of a cell differs from the life of the whole organism of which it forms a part ; for many cells may die and the life of the organism to which they belong not be affected. Important distinctions exist between vitality, life, and soul. A single cell may have vitality ; the individual organism to which the cell belongs has life ; and that organism, if possessed of self-consciousness, and of the power of self-direction, has soul. To assert Lotze's doctrine of an immaterial principle as the cause of form in organisms is not to assert the theory of vital force.

When I woke after my first night in Venice, which I had entered by the full moon, my earliest act was to ascend the tower of St. Mark's, and obtain a general view of the city by the rising sun. Before we discuss our central question, "Does death end all?" let us take a large view of this theme, as if from St. Mark's tower. Our rising sun here is the refulgent certainty that every change must have an adequate cause. When our national historian wrote the first volume of his history of the United States, it was not known that the Mound-builders had left elaborate traces of themselves in the spacious West. George Bancroft, therefore, asserted that the Mississippi valley was without any remains of human works. But since he wrote that first volume of his, we have discovered the most intricate kinds of mounds in the prairies ; and it is now universally conceded that there was a race of Mound-builders, and that the Mississippi valley is full of their works. On the prairie near Adrian, Michigan, for example, there is a night-hawk traced by mounds on the earth ; and the spread of its wings is two or three hundred feet. Over against him on the verdant, ancient acres, the mounds present the figure of a warrior with a balanced spear. Bancroft knew something of these mounds at the time he wrote his book ; but he said they were produced by geological action. In the Drift period these peculiar formations had been made by the complex swirls of the water and icebergs. If a man should undertake to hold to that theory now, and affirm that the Drift period formed these mounds, what would you say to him ? There is your night-hawk. Is it not possible for a complexity of geological forces—gravitation, chemical action, and the turmoil of a cooling planet, of which Strauss, Virchow, Häckel, and Huxley make so much—to trace on the prairie a night-hawk ? Is it not, at least, possible that your night-hawk might have been traced there by the movements of matter having in it the power and potency of all life? May it not be that thus were produced your savage and his balanced spear? You would say that a man holding such views ought to be sent to the lunatic wards. No *may be* is good for anything in science, unless it may be an *is*. But how about your actually living night-hawk, flying there above the prairie in the edge of the evening? How about your savage there miraculously alive, and poising his spear? Although you believe this rude earthwork tracery of the night-hawk and the savage cannot possibly have originated in any complexity of merely physical forces in a cooling planet, you will allow a man, if he is full enough of scientific authority, to come before you, and seriously puzzle you, as Strauss, Huxley, Virchow, and Häckel attempt to do, with the assertion that the bioplast—which stands at the head of the development of your living night-hawk, and which had in it all that has followed of life on this globe—came into existence in some Drift period by a fortuitous concourse of atoms. You ought for this to be sent to the lunatic wards. The reply to all reasoning of that sort is simply this, that *merely physical forces do not act so*. As Agassiz used to say, "The products of merely physical forces are the same in all quarters of the globe, and during all time known to man; but the products of the forces that produce life are varied under the same circumstances. Between two such sets of forces there can be no causal or genetic connection."[1] The results of the forces that produce organisms differ in different periods, and therefore we cannot account for them by these invisible, blind, mechanical laws. If, on the prairie, the figure of your night-hawk was not traced by a complication of these forces, assuredly, in the name of all clear ideas, the first bioplast that came into existence, and the bioplasts that weave the night-hawk and savage, were not constructed by any such complication of physical forces, acting without design or choice.

Does death end all? The answer to that question depends on the reply to another: Is life the cause of organisation, or organisation the cause of life? Is the relation of the soul to the body that of harmony to the harp, or that of the harper to the harp?

What are the strategic points in the discussion of the origin of life?

1. Tyndall, Huxley, Bain, Drysdale, and Spencer himself, all admit that the actions of bioplasts cannot be explained by merely chemical properties or forces.

If I succeed in showing you that this concession is made by the materialistic school, you will be relieved from much distress cast on you by popular irresponsible scribblers and declaimers. In November 1875, Professor Tyndall quoted and adopted these words of Du-Bois Reymond, "It is absolutely and for ever inconceivable that a number of carbon, hydrogen, nitrogen, and oxygen atoms should be otherwise than indifferent as to their own position and motion, past, present, or future."[2] Tyndall adds in his own words, that "the con-

---

[1] Agassiz, Essay on Classification.
[2] See Preface to Tyndall's Fragments of Science. Also his article in the Fortnightly Review, November, 1875, p. 585. Also Dr. Charles Elam's art. on "Automatism and Evolution," Contemporary Review, September 1876, p. 539.

tinuity between molecular processes and the phenomena of consciousness is the rock upon which materialism must inevitably split whenever it pretends to be a complete philosophy of the human mind." That is Tyndall, if you please, in 1875, writing a preface to the Belfast address, which needed much explanation after its errors had been searchingly pointed out by general public discussion.

There is inertia everywhere in all that we call matter. What is inertia? The incapacity to originate force or motion. Inertia is a property of the matter in bioplasm as surely as of that in any other part of the universe. This is the substance of DuBois Reymond's famous concession, that it is for ever inconceivable that a mass of physical atoms—past, present, or to come—should be outside the range of the law of inertia. "There is," says Faraday,[1] "one wonderful condition of matter, perhaps its only true indication, namely, inertia."

Even Herbert Spencer, who would be very glad to prove the opposite, says in his "Biology" (vol. i. p. 182), "The proximate chemical principles, or chemical units,—albumen, fibrine, gelatine, or the hypothetical proteine substance,—*cannot possess the property of forming the endlessly varied structures of animal forms.*" This is Herbert Spencer in 1864. "Nor," continues he, "can any such power be given to the cell as a morphological unit, even if it had a right to that title." It is the bioplast that is the morphological unit, and not the cell. "Therefore," concludes Spencer, "there is no alternative but to suppose that the chemical units combine into units immensely more complex than themselves, and that, in each organism, the physiological units produced by this further compounding of highly compound atoms have a more or less distinctive character. We must conclude, that, in each case, some slight difference of composition in these units, leading to some slight difference in their natural play of forces, produces a difference in the form which the aggregate of them assumes." Spencer's "Biology" is now an outgrown book, so rapid has been the progress of biological knowledge since its publication.

But the reply to this precious theory is, that *involution and evolution are a fixed equation.* If these multiplex molecules and their merely mechanical actions, which Spencer says build the body, have no life behind them, you will get no life out of them. If the smaller units out of which he makes up his larger units have no life in them, you will obtain from the latter only what was in the former.

Let us be for ever sure that the law of the persistence of force requires that evolution and involution should be equal to each other. You will get out of your molecular units what you put into them, and nothing more. But, according to Spencer himself, the chemical and physical forces and properties of atoms cannot build an organism. Larger molecular masses made up of these units, he says, may do so. Not unless there can be more evolved from, than is involved in, these units. If involution and evolution are *not* an eternal equation, there

---

[1] Correlation and Conservation of Forces, p. 24.

may be an effect without a cause. You cannot evolve anything which you have not first involved. Huxley, Spencer, Bain, and Drysdale, all admit that, if you make up your compounds from all the ascertained molecular activities, you involve nothing that will account for the weaving of these complex tissues. That admission is fatal to their further pretence, that a combination can be made which will evolve what has not been involved.

But Dr. Drysdale, who is a candid Scotch writer, makes a most distinct admission, that, even after we have built up these complicated molecular units, the matter in them must be *inert*. Hear the authority of a man who opposes Beale's opinion, that the action of the bioplasts cannot be accounted for except by a higher than physical cause, and who seriously undertakes, while admitting Beale's facts, to persuade the world that this matter in the bioplasts is of an infinitely peculiar sort, and that all it needs is "stimulus" to set it at work in all this miraculous weaving and inweaving and co-ordination of tissues. Dr. Drysdale says in so many words,[1] "*No matter how complex the protoplasmic molecule may be, its atoms are still nothing but matter, and must share its properties for good or evil, and among the rest inertia. Hence it cannot change its state of motion nor rest without the influence of some force from without. True spontaneity of movement is, therefore, just as impossible to it as to what we call dead matter. . . . So we are compelled to admit the existence of an exciting cause in the form of some force from without to give the initial impulse in all vital actions.* This is the—What? The soul? We expect him to say that; but what he says is, "This is the *stimulus*," whatever that may mean.

It is very surprising, in view of the school of thought to which Professor Alexander Bain of Aberdeen belongs, that, in his work on "The Senses and the Intellect" (p. 64), he should go so far as to uphold the doctrine of the spontaneity of vital actions, and to maintain that a spontaneous energy resides in the nerve-centres which gives them the power of initiating molecular movements *without any antecedent sensation from without*, or emotion from within, or any antecedent state of feeling whatever, or any stimulus extraneous to the moving apparatus itself. This fact of spontaneous energy he regards as the essential prelude to voluntary power.

So much, gentlemen, for the latest concessions of materialists; but I hold in my hand here the best, or certainly the freshest, book in the world on the "Cellular Theory;" and what are its opening words? All medical students in this audience will know that Professor Heinrich Frey of Zurich is a great authority on the cell-theory, and that this book of his has had an enormous sale between the Alps and the Baltic. Frey's work on "Microscopic Technology" is placed side by side with Stricker's "Histology" in the reading recommended to the two hundred young men in the Harvard Medical School yonder; but fresher than either of these books is this new volume published by Frey in 1875.

---

[1] Protoplasmic Theory of Life, p. 199.

Rufus Choate, as you remember, used sometimes to lay out a course of study in the classics perfectly parallel with that of the young men in Harvard University, and, in his breathless profession, would keep pace with them year after year. What if a student of religious science, who has no right to know anything about physiology, should look at the text-books in use in Harvard Medical School on physiology and other topics, and by this means, and by considerable conversation with men of science, assuring himself that he is not reading rubbish, and with a professional medical library at his command, should follow side by side the investigations those highly privileged young men are pursuing yonder, and occasionally stand with them in their dissecting-rooms? I know at least one student of religious science who does precisely that, and is fascinated with his work. Biology is now quite as interesting as the classics. In your Johns Hopkins University in Baltimore, studies are elective; and about ninety out of one hundred of the students there elect biology as one of their subjects.

Professor Frey of Zurich, in this work, which is hardly dry from the press, prints, face to face with the world, these as his very first sentences: " A deep abyss separates the inorganic from the organic, the inanimate from the animate. The rock-crystal on the one side, vegetable and animal on the other: how infinitely different the image! Is it, then, possible to bridge over this gulf? We answer, Not at the present time." We turn on in this volume, and find that reference is made to the theory that vital transformations are much like crystallisation, and that then these remarks are made, with a very apparent and not undeserved sly smile:

" Schwann, the founder of modern histology, taught, What the crystal is in regard to the inorganic, that the cell is in the sphere of life. As the former shoots from the mother lye, so, also, in a suitable animal fluid, are developed the constituents of the cell, nucleolus, nucleus, covering, and cell contents. *This view was embraced during many years, it explained everything so conveniently. This was, however, over-hasty.* The cell arises from the cell. *A spontaneous origin does not occur.*"[1] All this is in accord with what Huxley says in his article in "The Encyclopædia Britannica," " There is no parallel between the actions of matter in the mineral world and in living tissues."

2. After the unanimity of experts, there is no higher authority on any scientific doctrine than to find it taught in standard text-books in schools of the first rank; but you may easily ascertain that the very latest standard text-books oppose the mechanical or materialistic theory of life.

Dr. Tyson's book on " The Cell Doctrine " is in use side by side with Frey in your Harvard Medical School; but Tyson opens with diagrams from Beale, and closes with Beale; and where is there anything in him that is regarded as invulnerable, that he did not

---

[1] Professor Heinrich Frey, Compendium of Histology, Twenty-four lectures. Translated by Dr. G. R. Cutter. New York: Putnam's Sons, 1876. Pp. 1, 14.

obtain from Beale? Over and over, in the latter half of the book, as he closes the history of the thirty-nine years since the cell-theory was promulgated, he cites Beale; and, in spite of all the sneers from Huxley and others about "aquosity and horology," he sums up established science thus, "*We believe that the proper shaping, arrangement, and function of these elementary parts, is not a process identical or analogous to crystallisation, taking place through merely physical laws, but that there is a presiding agency which controls such arrangement to a definite end.*"[1] This is a statement out of a text-book mentioned officially in the catalogue of Harvard University as in use in the best medical school of your nation; and here is the best German book; and I have just read to you out of the best Scotch book; and Beale's is the best English Book; and they are all explicitly agreed in the assertion, that it is life, not mechanism, which weaves us and all things that live.

3. I affirm that we have under the microscope ocular demonstration that it is life which causes organisation, and not organisation which causes life. What is the first thing that appears in the formation of an organisation? A mass of germinal matter that has life but *no organisation.* You know what a naked bioplast is,—a little speck of glue-like matter, transparent, colourless, and, under the highest powers of the microscope and every other test known to man, showing no organisation, but yet capable of multiplex movements,—all these in a minute [referring to coloured diagrams on the platform]. "*We fail,*" Huxley says, "*to detect any organisation in the bioplasmic mass;*" but there are movements in it and life. We see the movements: they must have a cause. The cause of the movements must exist before the movements. *The life is there before organisation.* But, if life may exist before organisation, it may do so after it, or outside it.

If, according to custom in some rude games of sailors, we were to put a man in a canvas bag, and throw him in the bag upon this platform; and if that bag were to begin to cast out a promontory here, and a promontory there, and assume scores of shapes, and move to and fro, and pick up, now this object, and now that,—we should have no unfit representation of a portion of the movements of a naked bioplasmic mass. Your astonishing bag here picks up this chair, which cannot move of itself; and to make the parallel complete, it must have the power of absorbing this inanimate object, and of changing it into something just like itself, or alive. Suddenly this man in the bag may, if the parallel is to be made perfect, throw off a small sack from the bag, and that instantly begins to move on this platform: it forthwith commences to pick up lifeless matter, and to transform it into living matter like itself. It, too, throws off other little sacks, which go through the same motions again. We should say that sacks of that sort had very complicated machinery in them. But this is by no means the chief marvel.

You know, gentlemen, that in India it is a play of the children

---

[1] Dr. James Tyson, The Cell Doctrine, pp. 112 and 113. Lindsay & Blakiston, 1870.

and of grown men to make up the form of an elephant by stacking themselves together, two men making a leg of the elephant, six or eight his body, three or four his head, one or two his proboscis. You see in the pictures from India representations of elephants, made up, as you notice when you look at them sharply, wholly of human forms. Now, to carry out this parallel, we must have our first canvas bag transform itself into many canvas bags, and then all of them build themselves up, after this Indian fashion, into the elephant, the lion, the giraffe, or the palm-tree, the date, or the pomegranate; and these must live. They must *grow.* Some of the miraculous sacks will drop away from day to day; but new ones must take their places, and fill out the design had in view at the first. Of course, the part assigned to the man in the proboscis of an elephant thus built must be very different from that assigned to a man in the leg. If an elephant is to be made up in that way, the men who form his back must have a very different position from the men who form the tusks. There must be very peculiar activities put forth by each man in each part of your elephant. So, although our bioplasm is, to all appearance, the same thing when it weaves a tendon, and when it weaves a muscle, and when it weaves a nerve, its activities are very different. Surely the invisible molecular machinery must be very complicated indeed; for it makes a tendon here, a muscle here, or a nerve here. According to Spencer and this astute materialistic school, the bioplasts are nothing but molecular machinery, started off by "stimulus" into all this weaving, as the spark starts off the gunpowder into explosion. We say, that, if that is so, the molecular machinery must be more than exceedingly complex; for not only must it really be very different when it weaves a nerve from what it is when it weaves a muscle; but,—and this is the point on which to fasten supreme attention,—when we run back the examination of all our co-ordinated tissues, we find that assuredly all this molecular machinery must in some way have existed, or have been provided for, in the first little transparent, colourless, and apparently structureless bioplast which began to weave your elephant or your man, your pomegranate or your palm. A rather complicated kind of molecular machinery to be crowded into a space so small!

The acorn which hangs above the nest of your eagle has in it bioplasts that differ under the microscope in no particular from the little mass of bioplasm in the eagle's egg. Your bioplasm that weaves your oak is, to all human investigation, the same thing with the speck of bioplasm which weaves your eagle. Gentlemen, there is no inductive evidence of the existence of this mechanism. We may say, therefore, that, in the present state of knowledge, we cannot prove that molecular mechanism, acted upon by physical and chemical forces, is the sole source of organisation.

4. Matter in living tissues is directed, controlled, arranged, so as to subserve the most varied and complex purposes.

Only matter and mind exist in the universe.

Matter in living tissues must therefore be arranged either by matter or by mind.

No material properties or forces are known to be capable of producing the arrangements which exist in living tissue.

In the present state of knowledge, these arrangements must be referred to mind or life as their source.

5. Bioplasm exhibits peculiar actions found nowhere in not-living matter.

It exhibits different actions in every different animal and vegetable tissue.

For each class of these peculiar actions, there must be a peculiar cause.

That cause must be either matter or mind.

But the cause has qualities which cannot, without self-contradiction, be attributed to inert matter.

It must therefore exist in the life, or an immaterial element of the organisation.

6. It is plain that, before the matter which forms the tissues has entered the organisation, the plan of the tissues is involved in the earliest bioplasts.

There is forecast involved, therefore, in the action of the bioplasts. "Bioplasm prepares for far-off events," says Professor Lionel Beale over and over.

Forecast is not an attribute of matter, but of mind. An immaterial element exists, therefore, in living organisms.

7. There is a great fact known to us more certainly than the existence of matter: it is the unity of consciousness. I know that I exist, and that I am one. Hermann Lotze's supreme argument against materialism is the unity of consciousness. I know that I am *I*, and not *you;* and I know *this* to my very finger-tips. That finger is a part of my organism, not of yours. To the last extremity of every nerve, I know that I am one. The unity of consciousness is a fact known to us by much better evidence than the existence of matter. I am a natural realist in philosophy, if I may use a technical term : I believe in the existence of both matter and mind. There are two things in the universe ; but I know the existence of mind better than I know the existence of matter. Sometimes in dreams we fall down precipices, and awake, and find that the gnarled savage rocks had no existence. But we touched them ; we felt them ; we were bruised by them. Who knows but that some day we may wake, and find that all matter is merely a dream? Even if we do that, it will yet remain true that I am I. There is more support for idealism than for materialism ; but there is no sufficient support for either. If we are to reverence all, and not merely a fraction, of the list of axiomatic or self-evident truths, if we are not to play fast and loose with the intuitions which are the eternal tests of verity, we shall believe in the existence of both matter and mind. Hermann Lotze holds that the unity of consciousness is a fact absolutely incontrovertible and absolutely inexplicable on the theory that our bodies are woven by a complex of physical arrangements and physical forces, having no co-ordinating presiding power over them all. *I know that there is a co-ordinating presiding power somewhere in me. I am I. I am one.*

Whence the sense of a unity of consciousness, if we are made up, according to Spencer's idea, or Huxley's, of infinitely multiplex molecular mechanisms ? We have the idea of a presiding power that makes each man one individuality from top to toe. How do we get it ? It must have a sufficient cause. To this hour, no man has explained the unity of consciousness in consistency with the mechanical theory of life.[1]

There is not in Germany to-day, except Häckel, a single professor of real eminence who teaches philosophical materialism.[2] The eloquent Michelet, the life-long friend and disciple of Hegel, lectured at Berlin University in the spring of 1874 in defence of the Hegelian philosophy *as a system*. Out of nearly three thousand students he obtained only nine hearers. Helmholtz, the renowned physicist of Berlin, has come out through physiology and mathematical physics into metaphysics ; and his views in the latter science are pretty nearly those of Immanuel Kant. Wundt, the greatest of the physiologists of Heidelberg University, which leads Germany in medical science, has made for years a profound study of the inter-relation of matter and mind ; and he rejects materialism as in conflict with self-evident, axiomatic truth. Hermann Lotze, now commonly regarded as the greatest philosopher of the most intellectual of the nations, and who has left his mark on every scholar in Germany under forty years of age, is everywhere renowned for his physiological as well as for his metaphysical knowledge, and as an opponent of the mechanical theory of life.

I look up to the highest summits of science, and I reverence properly, I hope, all that is established by the scientific method ; but when I lift my gaze to the very uppermost pinnacles of the mount of established truth, I find standing there, not Häckel, nor Spencer, but Helmholtz of Berlin, and Wundt of Heidelberg, and Hermann Lotze of Göttingen, physiologists as well as metaphysicians all ; and they, as free investigators of the relations between matter and mind, are all on their knees before a living God. Am I to stand here in Boston, and be told that there is no authority in philosophy beyond the Thames ? Is the outlook of this cultured audience, in heaven's name, to be limited by the North Sea ? The English we revere ; but Professor Gray says that there is something in their temperament that leads to materialism. England, green England ! Sour, sad, stout *skies*, with azure tender as heaven, omnipresent, but not often visible behind the clouds ; sour, sad, stout *people*, with azure tender as heaven, and omnipresent, but not often visible behind the vapours. Such is England, such the English. We are to extend our field of vision to the Rhine, to the Elbe, to the Oder, to the Ural Mountains ; and, when we look around the whole horizon of culture, the truth is, that philosophical materialism to-day is a waning cause. It is a crescent of the old moon ; and, in the same sky where it lingers as a ghost, the sun is rising, with God behind it.

---

[1] See Lotze's greatest work, Mikrokosmus, Leipzig, 1869. Vol. i. book 3, chap. 1.
[2] See art. on "Philosophy and Science in Germany," Princeton Review, October 1876, pp. 752–755.

VIII.
DOES DEATH END ALL? THE NERVES AND THE SOUL.[1]

"It needs not that I swear by the sunset redness,
And by the night and its gatherings,
And by the moon when at her full,
That from state to state ye shall be surely carried onward."
KORAN.

"Die Kraft, die in mir denkt und wirkt, ist ihrer Natur nach eine so ewige Kraft, als jene, die Sonnen und Sterne zuzammenhält. Ihre Natur ist ewig, wie der Verstand Gottes, und die Stützen meines Daseins—nicht meiner körperlichen Ercheinung—sind fest, als die Pfeiler des Weltalls."—HERDER, *Philosophy of History*.

PRELUDE OF CURRENT EVENTS.

SAFE popular freedom consists of four things, and cannot be compounded out of any three of the four—the diffusion of liberty, the diffusion of intelligence, the diffusion of property, and the diffusion of conscientiousness. In the latter work, the Church is the chief agent; and her most important instrumentality we call the Sabbath. Goldwin Smith very subtly says that it is free religion and hallowed Sundays which explain the average moral prosperity of America. We have had in the last week, in Boston, a somewhat obscure and erratic convention, advising America to do better than she has thus far done in following the New-England ideas concerning Sunday. Give America, from sea to sea, the Parisian Sunday, and in two hundred years all our greatest cities will be politically under the heels of the featherheads, the roughs, the sneaks, and the moneygripes. Abolish Sunday, and the social sanity it fosters, and, in less than a century, the conflict between labour and capital would issue here in petroleum fire-bottles. Capital in our great muncipalities is fleeced now to the skin. Does it wish such social insanity to spring up as shall cut it through the cellular integument to the quick? If it does, let capital abolish Sunday. Working-men desire to build co-operation up into a palace for themselves and their little ones; and God speed their effort to protect their own! But how can co-operation succeed without the large confidence of man in man? and how can that come without the moral culture given by the right use of Sundays? Co-operation fails because men are not honest. How are men to be made honest without a time set apart for religious culture? That population which habitually neglects the pulpit, or its equivalent, one day in seven, can ultimately be led by charlatans, and will be.

---

[1] The fifty-third lecture in the Boston Monday Lectureship, delivered in Tremont Temple.

I am no fanatic, I hope, as to Sunday; but I look abroad over the map of popular freedom in the world, and it does not seem to me accidental that Switzerland, Scotland, England, and the United States, the countries which best observe Sunday, constitute almost the entire map of safe popular government.

Sabbath is a day of religious culture and cheerful rest. Its biblical warrant is found in the re-affirmation by the Sermon on the Mount of the whole moral spirit of the Decalogue. I affirm, without fear of successful contradiction by any cultured thought, that the Sermon on the Mount re-affirms the moral spirit of the Decalogue, and in that re-affirmation perpetuates the direction to hallow one-seventh portion of our time: it matters very little which seventh. "Forsake not the assembling of yourselves together," is apostolic precept, as it was apostolic example. No doubt small critics may show that the apostles and our Lord did works of necessity and mercy on the Sabbath; and so do we, and so will we to the end of time. But the Sermon on the Mount re-affirms your first, your second, your third, your fifth, sixth, seventh, eighth, ninth, and tenth commandments. How are you to show that it does not re-affirm the fourth in spirit? "Not one jot or tittle shall ever pass from the law till all be fulfilled."

It is fifteen hundred years now since Constantine put into execution the law bringing one day in seven an unwonted hush on all industry in the Roman dominion. Here we are ten centuries off from the time when Christianity closed her chief political struggles. Here is a republic built chiefly by Christianity, and perfectly free, and governing more square miles than ever Cæsar ruled over. This nation calls peace to her industries one day in seven. She sends nine millions of her population, one in five, to a World's Fair, and shuts the door every Sunday. I know what report says about the evasions and hypocrisy of the Centennial Commission in admitting persons surreptitiously into the buildings on the Sabbath against the vote to close the grounds on that day. If the report is correct, the Centennial Commission ought to have public rebuke, unless it can make adequate explanation.

I am glad to see that even this erratic convention, dazzled out of sight by the sound ideas and majestic words of the Episcopal congress, was wise enough to proclaim that it did not wish to introduce into America the European Sunday.

Hallam says that European despotic rulers have cultivated, as Charles II. did in the day of the "Book of Sports," a love of pastime on Sabbaths, in order that their people might be more quiet under political distresses. "A holiday Sabbath is the ally of despotism." Wherever the Romish or Parisian Sunday has prevailed for generations, it has made the whole lives of peasant populations a prolonged childhood.

America, I venture to say, is satisfied with the record of the Sabbaths in her World's Exhibition. This convention seemed to think, however, that the burden of a great reform was laid upon its shoulders. It apparently thought its thin meetings the representation of a large constituency. Men are strangely full of company sometimes, when before the mirrors of high self-appreciation. Sidney Smith, calling on a nobleman, passed through a room full of mirrors, which showed him several images of his own form approaching from many directions. He was wholly alone; but he was overheard to say, "A meeting of the clergy, I see."

THE LECTURE.

Suppose that the musician at your organ yonder has on his finger Gyges' ring, which according to the Greek mythology, as you remember, made the wearer invisible. It is entirely clear, is it not, that if we were to approach and study that instrument while it is in action under the fingers of this invisible musician, we should find in it no authority for attributing the anthem proceeding from the organ to the inert matter composing the organ? We should have, on the contrary, incontrovertible evidence in the very structure of the instrument that it was made to be operated upon from without. If it is to give forth melody, it must be moved by something not itself. It is composed of wood and metal and ivory, all of which, with all their complicated mechanical arrangements, are inert, and, if taken alone, are wholly valueless in the production of music.

In one portion of the organ we have a keyboard ; and, in the case supposed, we look on the very intricate combinations and motions in the keys, and see no cause for the movements. But we know, if we are sane, that every change must have an adequate cause. We find a perfect correspondence between the motions of the keys and the pulsations of the melody rising and falling in this temple. But this parallelism is not identity. The keys in motion are not the music. Motions and forces are not the same.

Let, now, some inquirer of narrow mental horizon, and confusing —as so much current discussion does—motions with forces, assert that these intelligent movements of the keys—which, of course, must have behind them forces containing intelligence—are the sole cause of the anthem. *Let him insist on a new definition of ivory.* Let him affirm that the matter composing these keys has in it the power and potency of all music, from the simplest air up to Beethoven's Fifth Symphony. Let him go behind the organ, and elaborately study the very powerful and purely physical forces at work in the interior of the instrument. Let him show, learnedly and laboriously, that currents of air thrown into the pipes produce, according to merely mechanical principles, the wholly physical concussions in the molecular particles of the atmosphere which are concerned in the music. As no merely physical science, by any test known to man, can detect the presence of the musician, let this observer assert that there is no musician independent of the instrument, and that the anthem proceeds wholly from the mechanism of the organ, acted upon by exclusively physical stimulation from without. Let him assert that the hypothesis of an invisible musician is as absurd as the attribution of aquosity to water, or of horology to a clock. According to this supposed materialistic observer of the organ, there is nothing in the anthem which is not wholly the result of the mechanism of the organ on the one hand, and of the merely physical forces supplied to it by the organ-bellows on the other. Let this naturalistic observer have a great name—among men of his own opinions.

Should we be puzzled by these confident assertions? Not if we held fast to the Ariadne clew of the self-evident, axiomatic truth, that every change must have an adequate cause. We should say that this instrument, being made wholly of matter, is inert. We should assert, in the name of established science, the incontrovertible inertness of all parts of the organ taken alone. We should say that the motion of rough currents of air through it does not and cannot account for the intricate and ravishing melody which captivates our souls by its intelligence, and must have behind it a soul. Mere wood, metal, and ivory cannot utter Beethoven's spirit. Perhaps the air, by the slight pressure of intelligence on the keys, can be ruled into melody, and made to give all its majestic force to the intelligent weaving of the anthem. But in your organ, as elsewhere, involution and evolution are a fixed equation. You bring out of it only what you put in. Your musical instruments will throw no Beethoven into the air, unless there is a Beethoven at the keys.

Such, my friends, is the stern outline of the ineffaceable contrast between the body and the soul. The distinction between matter and mind is a gulf as vast and impassable in physics as in metaphysics. The soul wears Gyges' ring. It is, indeed, invisible to the microscope, and intangible to the scalpel. But there are mysterious molecular motions in the nervous substance of the brain. Neural tremors fill the keyboard of the body. Undoubtedly there is a perfect correspondence between these tremors and the anthems of thought and emotion, in your Homer, your Demosthenes, your Cæsar, your Milton, your Shakspeare. But the parallelism is not identity. Motions and forces are not the same. The keys in motion are not the music. Physical forces play through the brain; but they do not sing, unless modulated by the ineffable touches of the keys. Just as surely as you, from the structure of an organ, may infer the necessity of a wholly exterior agent to move it, so, from the structure of the nervous system, we must infer the necessity of a wholly external agent to set it in action.

In what I am about to put before you I have the authority of Frey, of Stricker, of Ranke, of Kölliker, of Carpenter, of Beale, of Dalton, and of Draper.

1. In the nervous mechanism there are two kinds of fibres, called by physiologists the automatic arcs, and the influential arcs.

We have here a representation of the simplest kind of nervous fibre [illustrating by a figure upon the blackboard],—the pendent curve of a nervous thread, one end in contact with the external surface of the body, and the other connected with this muscular tissue. If you please, the bioplasts weave all that. Perfectly simple as the structure looks, it is a miracle. Can you make anything like it? Here is your muscular fibre, which has the peculiar quality of contracting under nervous stimulus. Here is your nervous cord, which transmits strange influences that cause contraction when they are received upon this muscular tissue. One test by which true is to be distinguished from false science is, that the former does, and that the latter does not, concern itself carefully with beginnings. Remember, that, even in this automatic nerve, motions and forces are not the same. Muscular contraction is an effect of physical forces only as these act on mechanism arranged before the forces themselves came into play. Your miraculous brain is first woven by your bioplasts. You say mind is the result of the mechanism of the brain; but the mechanism of the brain is the direct product of bioplasmic action.

Of course, I am ready to admit, that, if you touch a portion of this automatic nervous arc with a galvanic current, you will produce contraction there in the attached muscle. Electrical stimulation of such a nerve may produce a contraction of the muscle even after the man is dead. But what wove that nerve? What wove that contractile tissue?

Beyond this simplest structure, the next higher in the development of the nervous system is what is called the cellated nervous arc. We see it here, a pendent curve as before ; but now with a very large bead, or mass of nervous matter with bioplasts in the middle of it, is hanging at this point. It is yet true that irritation here produces contraction there. What influence, then, has this nervous centre upon the transmission of this nervous force? The books say that there is no proof that the nervous influence is changed in quality by its passage through one of these simplest ganglia. You may single out a nerve arc of that primitive style, and irritate it by an electric current on one side of this large bead or ganglion, and you will produce contraction in the muscle just as before. You irritate this side beyond the great bead, and you produce contraction.

But a third step in the development of the nervous system does introduce a change. Many of these nerve-centres are tied up to other nerve-centres [illustrating by a figure in which the ganglion of the nerve-arc was connected with another ganglion]; and in a nerve with its ganglion connected in that style with another ganglion, a portion of the influence transmitted through this complex nervous mass is thrown off into this other complex nervous mass. Your physiological authorities call the latter a registering ganglion. This transmission of nervous influence into the registering complex of nervous matter may be very inadequately illustrated, Professor Draper says, by a faucet with three stops,[1] or by a mirror with a portion of the isinglass taken off the back. The light is in part reflected and in part transmitted. Thus this registering mass of nervous matter retains a portion of the force sent through this nervous arc ; and, in an animal possessing this nervous mechanism, there will be memory, or something equivalent to it.

Thus far we have seen only what is called the automatic nervous mechanism. Please fix in your minds, gentlemen, the simplicity of this structure, and, when a more complicated mechanism is outlined in connection with this, keep vividly before your minds the contrast between the two.

All established science is agreed that there are automatic and also influential arcs in the nervous system, and that the contrast between the two things is as marked as that between their accepted scientific names.

In the higher animals there is added to the simpler automatic part of the nervous system a far more intricate structure, called the influential nervous mechanism. Professor Draper represents the contrast between the automatic and the influential part of the nervous system by this ideal figure,[2] which I here reproduce line for line. It is substantially a lower curve and an upper curve—the one automatic, the other influential, and the two bound together by nervous threads. In all physiology, outside the supreme topic of

---

[1] Professor J. W. Draper, Human Physiology, p. 380.   [2] Ibid., p. 282.

bioplasm, I know nothing which is so suggestive as this contrast between the automatic and the influential nerve-arcs. Here, assuredly, is a majestic mount of vision upon which the philosophy of the relations between body and soul, matter and mind, must often pace to and fro.

2. Plants and many animals possess only the automatic arcs.

3. Such organisations as possess only the automatic arcs are automata; and, although they have life, they cannot, in the strict sense of the word, be said to possess souls including free-will and conscience.

The contrast between the influential and the automatic is that between freedom and necessity. It is that between man, with the power of choice, and your poor honey-bee, who is supposed to work as an automaton. The bee has not the influential arc : it has only the automatic nerves. Accordingly, by instinct it has built its cell in the same way age after age. Two bees under precisely the same circumstances will do precisely the same things.

But this upper arc, which is possessed by man, is called influential, and not automatic, because it is the seat of activities of a free sort. This is the keyboard of your invisible musician : this is the white ivory shaped by no mortal fingers, and on which life plays.

Gentlemen, I have been accused of being rhetorical ; but a man who wishes to dazzle by rhetoric does not talk in twenty-eighthlies and forty-ninthlies, as I have sometimes done. Any one, however, who wishes to convince by cool precision, very naturally employs numerals. You will allow me, therefore, to number the points of a discussion, which must be crowded, and which would nevertheless be clear.

Just here expose themselves in more than glimpses the fascinating questions as to the difference between instinct and reason, and as to the immortality of instinct. Animals that possess only the automatic nerve-arcs have only instinct for their guidance : they have life, but not free wills and consciences. Later in this course of lectures, I shall discuss the question, whether, after death, there is a survival of the immaterial principle in animals that are mere automata. Here and now I emphasise only this broad distinction between the influential and automatic nerve-arcs, a physical fact, without any haze either in its margin or its contents. God materialises. In the universe of forms, as well as in that of forces, the Divine language has no empty syllable. Perhaps this invisible musician, with Gyges' ring on his finger, has not been left without a witness of himself in the whitish-gray keyboard of the human organ. Perhaps the contrast between the automatic and influential nerve-arcs is just as important a fact in the instrument God has made as the distinction between your musician and the man who moves the bellows behind the organ is in the instrument man has made. Among the automatic and influential nerve-arcs, all philosophy ought to stand listening with hushed breath.

4. Man possesses in abundance both the automatic and influential arcs.

5. Whatever animal possesses the influential arcs has a depository, magazine, or reservoir of force not dependent on external impressions.

Aristotle noticed with great keenness of interest the fact that men awake before they open their eyes. Professor Bain regards that circumstance, with which we are all familiar, as one out of thousands of proofs that external irritation is not necessary always to internal activity.

By the way, Aristotle was accustomed to assert that the most interesting portion of human knowledge is that which refers to what he called the animating principle of physical organisms. We are beginning to think, I hope, that what is called bioplasm is the most interesting by far of all the objects known to physical science. That, in substance, is an opinion two thousand years old. Aristotle defined the animating principle as the *cause of form in organisms*.[1] This to him was the most alluring of all the topics open to Greek philosophy. He said often, that, if we ought to be interested in a theme in proportion to its dignity, certainly nothing could be more entrancing than the study of the animating principle.

6. In man the influential arc is the seat of intellect, free-will, and conscience.

7. But, as man possesses the automatic arc also, many of his actions are automatic.

We must expect to find in some animals which have a much more perfect automatic nervous mechanism than man, instincts, and, apparently, spontaneous movements, of the most marvellous kinds. I am not asserting that man is not in some respects an automaton; but he is by no means as good a one as might be chosen if the power of automatic nervous action is to be shown. Professor Huxley went before a great audience at the Belfast meeting of the British Association for the Advancement of Science, and took a headless frog, and put it on the back of his hand, and then turned his hand slowly over; and the frog kept his place till the hand had been reversed, and the frog stood in the palm.[2] Now, said Professor Huxley, is there any will concerned in that? Is not this the result of purely physical stimulation of the frog's nerves? Have we not here an automaton? He meant to puzzle the world about the freedom of the human soul. But the bioplasts wove that frog too. After the automatic mechanism is woven, such results are very well known to follow the action of the merely automatic part of the nervous system. A frog with its head cut off you may put on the back of your hand, and you may turn the hand over, and the frog will keep its place meanwhile without assistance, and stand on your palm. Of course, there is no action of the cerebral hemispheres there. The irritation of the feet has such an effect as to cause the muscles to enable them to cling to their support; just as, while the perching bird sleeps, the perch itself stimulates to action the muscles that cause it to be clasped by the

---

[1] Aristotle de Anima, *passim.*
[2] Huxley's Address on the Question, Are Animals Automata?

bird's feet. Will you please notice that you have no right to be puzzled by any number of facts like these, and that all there is in Huxley's famous experiment is admitted truth concerning the automatic part of the nervous system, and that the puzzle consists in putting that fragment for the whole?

8. As in man, the automatic and the influential nervous arcs are blended together by innumerable commissures, and are yet perfectly distinguishable by study, so the automatic and the free activities of man are, in experience, most intricately blended together, and yet are perfectly distinguishable by careful attention.

9. Sometimes the former may become so powerful as to overcome the latter; and sometimes the latter may overcome the former.

10. The power of habit, and, to a great extent, that of emotion, depends on the action of the automatic arcs.

Your classical orator of Boston stands upon some transfigured platform, and the warp and woof of his unpremeditated language fall from the loom of his mind, every figure perfect. You hold up in print the next morning his speech between your eyes and the merciless sunlight, and there is no flaw in the weaving. Your Phillips, your Everett, your Sumner, your Webster, have scarred into their nervous systems good literary habits. You know very well that a scar will not wash out, or grow out. Absolutely there is no doubt about this. But how vast and fathomlessly practical are the applications of the simple truth that scars are ineraseable! A two-edged sword this, and of keener than Damascus steel. Your dull inebriate, who scars his brain by the habit of intemperance, thinks, that, after his reformation, his nervous system will slowly recover all the soundness it once had. But in your finger a scar will not grow out; and on your brain a scar will not grow out. Here are scars which were made when my fingers were too young to be trusted with edged tools; but, although the particles of my body have been changed many times since then, the scars are here, reproduced with the reproduction of the particles of the body. Once in seven years we have a new body, the books used to say: once in twelve months, as they say now, the particles of our physical system are changed. Scars, however, are absolutely unchangeable in the changing flesh. We carry into our graves the marks of boyhood's sports; and this is as true, if you please, of the sports that scar the brain as of those that gash the fingers. The most searching blessing on good habits, the most penetrating curse on bad, is found in the one fact, that the automatic nervous mechanism is such, that when a habit, good or bad, is scarred into the nerves and brain, the soul pours forth the result of the habit almost spontaneously.

The influential nerve-arcs can, indeed, hold back the activity of the automatic arcs. "The will counts for something as a cause," says Huxley himself. Dr. Carpenter explicitly teaches, that the influential nerve-arcs may resist, "keep in check and modify" the action of the automatic nervous mechanism.[1]

---

[1] Carpenter, Physiology, eighth edition, 1875, p. 730. See, also, his Mental Physiology, *passim.*

The power of volition resides in the influential arcs. But even a man is so far an automaton that, if he is an orator, he will scar himself with the complete oratorical habit, and may speak, as the bird sings, without effort. You wonder at the precision, fluency, and force of the language of your Burke or your Chatham. But the automatic nerve-scars representing good literary habits may have been in the mother, or in both parents, or in five generations. Certainly the habit of good extemporaneous speech has been cultivated through more than a quarter of a century by your Chatham and your Burke. It is now scarred deeply into the nerves; and scars do not grow out. And when, before any audience, the warp and woof of eloquent speech are needed, the automatic action of good habit sets its power behind the will of the orator; and nearly all that is required is, that some great thought and passion should throw the shuttles once, and then the figured, firm web flows spontaneously from the perfect loom. But just so, my friends, your tendency or mine to slovenly speech, our fearfully unæsthetic ways, and even the inebriate's thirst, or the sensualist's leprous thoughts, scar the nervous system in its automatic arc. When you, thus scarred by habit, and it may be, alas! by inheritance, pass the place of temptation, you are seized, you know not with what power: you feel that there is necessity upon you; and that mystery is simply the fact that scars are ineraseable. You have scarred your nervous system with an evil habit; and now this terrific power of the automatic mechanism stands behind your will. Your musician yonder, under the same automatic law, derives power from the very source from which you derive weakness. He calls forth melody, spray after spray of the fountain of the anthem ascending and falling, with raptures all in rhythm; and we are lifted by it to the azure; we are ennobled by it mysteriously: but your musician is making no effort. So has habit ingrained his nervous mechanism, that he plays as the bird sings. Professor Huxley states, that once an old soldier, who had been accustomed all his life to come to a perfectly erect attitude at the word "attention," was carrying home his dinner on a London street, when a comrade who desired sport called out to him from the other side of the way, "Attention!" Instantly the inattentive soldier came into the upright attitude, and dropped his dinner in the street. Now, Professor Huxley says, that although the details of that anecdote may not be all correct, they might be, and that they might be because of the power of the automatic action of the nervous system. So you, holding your families' or your own pure character in your arms; you, citizens of Boston, holding your honour in this city in your bosoms, are some day tempted sorcerously by intemperance or passion, by the greed and fraud of crooked trade or politics, or by any of the bad impulses that habit or inheritance has woven into your nerves; and suddenly, under automatic trance, which might yet have been escaped by force of will, the things dearest to you are dropped by you in the draggled street of your private or public life at the sudden word "Attention" from the black angel.

11. The action of the influential arcs is not to be regarded as a

creation of force, but rather as the optional opening of a reservoir of force, given with the gift of life to each organisation that possesses free-will.

I touch here upon a great mystery, and am quite aware of the nature of the ground over which I pass ; but you will notice that this proposition does not go as far as Sir John Herschel does, when he asserts that the soul is, to a small extent, a real creative force. Let us call it, rather, a power delegated for optional use. All the power we have is certainly delegated power. We have received it all from Almighty God. His force is all the force there is in the universe, intellectual or physical.

12. This fact, that free-will is exercised through the influential arcs of the nervous system, does not, therefore, necessarily contradict the law of the persistence of force.

13. In both the automatic and the influential arc there is a perfect adaptation of the structure to the agent that is to set it in activity.

Sometimes, at the end of the automatic arc, you have an eye, with its marvellous lenses, or an ear, which Professor Tyndall calls "a harp of three thousand strings."

14. The eye is the outer portion of the automatic arc concerned in vision ; and all parts of the eye are adapted in their structure to a wholly external agent,—light.

15. The ear is the outer portion of the automatic arc concerned in hearing ; and it is adapted perfectly to an external agent,—sound.

16. The nerves of smell are connected with a structure adapted to a wholly external agent,—odour.

17. The tongue is adapted in the same way to a wholly external agent,—flavour.

18. Many problems in biology are susceptible of an inverse solution : as, for example, given the nature of light to determine what must be the structure of the organ of vision ; or, given the structure of the eye to determine what is the nature of light.

19. So, in relation to the agent which moves the influential arcs, we have the problem : Given the structure of the brain to determine the nature of the agent which sets it in action.

20. There is an absolute analogy in construction between the elementary arrangement of the fibres of the brain and those of any other nervous arc.

21. The influential, as well as the automatic part of the nervous system, has its centripetal and centrifugal fibres, which converge to sensory ganglia, or nervous centres.

22. Just as the automatic arcs in man's nervous system have vesicular material at their external extremities in the organs of the senses, so the influential have vesicular material at their external extremities in the convolutions of the brain.

23. But we know beyond question that the automatic nerve-arcs can display no phenomena of themselves : they all require an external agent to set them in motion.

24. The optical apparatus is inert without the influences of light ; the auditory inert without sound. The organs of taste and smell, and

the nerves connected with them, are inert and without value, except under the influences of wholly external agents.

25. Established science asserts the absolute inertness of the cerebral structure in itself; or the entire incapacity of the influential as well as of the automatic nerve-arcs to initiate their own activities.

26. As, therefore, from the structure of the eye, we may infer the existence of a wholly external agent, light, or from that of the ear, the existence of a wholly external agent, sound, so, because of the absolute inertness of the cerebral structure in itself, we must attribute its activities to an agent as external to it as sound is to the ear, or light to the eye.

27. That agent is invisible to the external vision, and intangible to external touch.

28. It is positively known to consciousness, or the internal vision and touch.

29. That agent is the soul.

30. As the dissolution of the eye does not destroy the light, the external agent which acts upon it; and as the dissolution of the ear does not destroy the pulsations of air, the external agent which acts upon it; so the dissolution of the brain does not destroy the soul, the external agent which sets it in motion.

Gentlemen, there is more than one soul here besides mine sad with unspeakable bereavement. There are eyes here besides mine which weary the heavens with beseeching glances for one vision of faces snatched from us in fiery chariots of pain. Is death the breaking of a flask in the sea? Is there for me no more personal immortality than for a consumed candle? Cool precision, gentlemen, not rhetoric; even at the edge of the tomb, cool precision!

I open Professor Draper, and read, "If the optical apparatus be inert, and without value save under the influence of light; if the auditory apparatus yields no result save under the impressions of sound,—since there is between these structures and the elementary structure of the cerebrum a perfect analogy, we are entitled to come to the same conclusion in this instance as in those, and, asserting the absolute inertness of the cerebral structure in itself, to impute the phenomena it displays to an agent as perfectly external to the body, and as independent of it, as are light and sound; and that agent is the soul."[1] That is a very sacred kind of Scripture, for it is the record of God's work fairly interpreted.

I might quote twenty other authorities; but I cite this book because it has a great fame in Germany, and is accessible to all, and because Professor Draper, in a most painfully unfair volume on "The Conflict between Science and Religion," has set himself somewhat outside the pale of what I call just sympathies in this great discussion. He, at least, has proved his freedom from all traditional opinions. The objection to the latter book is, that he confuses Romanism and Christianity, and shows that conflict has existed between some forms of the Church and science, and then infers that Christianity itself is

---

[1] Draper, Physiology, p. 285.

in conflict with clear ideas. This man, with more than one compeer of his in the latest physiological research seconding his words, affirms, in the face of the world, that "It is for the physiologist to assert and uphold the doctrine of the oneness, the accountability and the immortality of the soul, and the great truth that, as there is but one God in the universe, so there is but one spirit in man."[1] "We have established the existence of the intellectual principle as external to the body."[2] That is Beale, and that is Hermann Lotze, too.

There is a school of rather small philosophy in Cambridge yonder, among a few young men, who, very unjustly to Harvard, are supposed by large portions of the public to represent the University. I happen to be a Harvard man, if you please, and ought to know something of my *Alma Mater*. There is not a paving-stone or an elm-tree in Cambridge that is not a treasure to me. Who does represent Harvard? Hermann Lotze and Frey and Beale, rather than Herbert Spencer and Häckel, are the authorities which the strongest men at Cambridge revere. The North American Review, the Harvard chair of metaphysics, the Harvard pulpit, the Cambridge poets and men of letters, who are tall enough to be seen across the Atlantic and half a score of centuries, are not converts to materialism.

Must I infer that the New York *Nation* is possessed of a philosophy of materialistic tendency? I have not criticised, I have even defended, the theistic doctrine of evolution. I have endeavoured only to show that the atheistic and agnostic forms of that doctrine are violently unscientific. There is a use and an abuse of the theory; and Dana represents the one, and Häckel the other. I have treated atheism and materialism without much reverence; for I revere the scientific method. But three weeks in succession I am assailed with ridicule without argument in a critical journal that claims to be courteous and fair. As this cultured, and, I may say, distinguished Boston audience knows, the New York journal has stated my positions with the most broad and painful inaccuracy. Am I to stand here before an audience that has as much culture in it as any weekly gathering in the United States, and be lashed before the world by this New York weekly, which is, indeed, well informed in politics, but in philosophy is so far behind our times as to be now predominantly Spencerian? Its editor, as you know, resides in Cambridge; and the small, disowned school in philosophy there seems to have taken possession of this periodical of very unequal merit. In philosophy, the *Nation* has no outlook beyond the Straits of Dover. I do not remember that I ever saw in it a single reference to Hermann Lotze, or any proof of large knowledge of so much as the outlines of the freshest German thought of the first rank on the physiological side of metaphysical research. As to present culture in the wide and rich theological field, I may say, that, so far as a specialist's judgment is worth anything, mine is, that the *Nation* cannot be trusted on this theme, it is so benighted by its insular philosophy, and by a very frequent arrogance toward all theology not Spencerian.

---

[1] Draper, Physiology, p. 24.   [2] Ibid., p. 286.

This paper needs a rival. I dislike to criticise it; for, after all, it is our poor best in the way of a critical weekly. At a hotel table in Munich once, a haughty English lord asked me what was the best paper in America of the order of the Saturday Review of London. "The Nation," I said. "Yes," he replied; "but you have forty millions of people, and Great Britain has only thirty millions, and you have but one paper of this class."

There used to be a proverb that, when Philadelphia wanted to know what to think, she looked to New York; and when New York wanted to know what to think, she looked to Boston; and when Boston wanted to know what to think, she looked to Concord. No doubt this proverb originated in Concord. But I walked the other day with a Concord author whose words have been read with delight by two generations, and will be remembered, I hope, by twenty; and he said to me under those historic elms on your Boston mall, after having been twice in the audience of this Lectureship, "You may tell Boston that I, for one, regard Lionel Beale and Hermann Lotze as the rising men in philosophy." That is Bronson Alcott, who lives not far from the spot where Nathaniel Hawthorne lies at rest till the heavens be no more. If you listen to the inner voice of Emerson's latest publications, and to that of Carlyle's, you will find that these men whom you have called pantheists, are no deniers of the personal immortality of the soul.

Am I out of my field in endeavouring to prove that man has a soul? *Ne sutor ultra crepidam.* Let no shoemaker go beyond his last, Horace said ages ago. But what if, in the progress of the ages, there be made a new last? Significant signs of the times are the professorships and lectureships starting up in renowned theological schools on the relations between the religious and other sciences. In New York City, in Union Seminary, there is a lectureship, with ten thousand dollars endowment on "The Relations of the Bible to the Sciences." It is called the Morse Lectureship, because founded by Samuel F. B. Morse, in memory of his father, who was only a doctor of divinity. In the same school there is a lectureship on "Hygiene," founded by Willard Parker. We have the Vedder Lectureship at the New Brunswick School of the Reformed Church in America. Princeton has a chair, established in 1871, designed to discuss elaborately "The Relation of Christianity to Natural and Speculative Science." Andover has a lectureship, and I hope may soon have a professorship on this theme. Out of place! I maintain that all these foundations are timely, and deserve the cordial support of all scholars. They are a new last, indeed; but the occupants of these chairs will make specialists of themselves in their new fields, which will by no means be outside the range of theological research. All these facts were overlooked by the *Nation* when it made its astute examination of catalogues to see whether ministers know anything of the latest philosophy. Catalogues are a sufficiently sorry authority; but their less slovenly perusal might have taught this journal that a new last has been created by a new time, and that, in the name of Horace's maxim no student of religious science can be warned off the field which Hermann Lotze and Beale have entered. No student of religious science is adequately equipped for his work, unless, with open eyes, he has worshipped in that temple of physiological research where Lotze and Helmholtz and Frey and Wundt and Beale and Carpenter and Dana, and all men of science who think not to twenty only, but to thirty-two points of the compass, now kneel, hushed, dead, in the presence of a Living God, but ready to rise up alive and fill civilisation with their own enthusiasm.

IX.
DOES DEATH END ALL? IS INSTINCT IMMORTAL?[1]

"Des Todes rührendes Bild
Steht Nichts als Schrecken dem
Weisen, und nicht als Ende dem Frommen."
GOETHE, *Hermann und Dorothea.*

"Die Schöpfung hängt als Schleier, der aus Sonnen und Geistern gewebt ist, über dem Unendlichen, und die Ewigkeiten gehen vor dem Schleier vorbei, und ziehen ihn nicht weg von dem Glanze, den er verhüllet. . . . Ich und du, und alle Menschen und alle Engel und alle Würmchen ruhen an seiner Brust, und das brausende, schlagende Welten- und Sonnenmeer ist ein einziges Kind in seinem Arm."—JEAN PAUL RICHTER, *Hesperus.*

PRELUDE ON CURRENT EVENTS.

ON the morning of Saturday, Oct. 23, 1852, Daniel Webster, whose statue was unveiled last Saturday in Central Park, said to his physician, "I shall die to-night." Dr. Jeffries, much moved, replied, after a pause, "You are right, sir." The gorgeous and jewelled October day rolled on at the edge of the sea; and when evening came, the last will and testament of your greatest statesman and orator was brought to him for his signature, which he affixed, and then said, "Thank God for strength to do a sensible act! O God, I thank Thee for all Thy mercies." His family was brought to his bedside; and his biographer, Curtis, noticing that Mr. Webster was about to say something which should be recorded, took his seat at a table, and caught these last words. Curtis says they were uttered slowly in a tone which might have been heard through half the house: "My general wish on earth has been to do my Master's will. That there is a God, all must acknowledge. I see Him in all these wondrous works. Himself how wondrous! What would be the condition of any of us, if we had not the hope of immortality? What ground is there to rest upon but the gospel? There were scattered hopes of the immortality of the soul, especially among the Jews. The Jews believed in a spiritual origin of creation. The Romans never reached it; the Greeks never reached it. It is a tradition that communication was made to the Jews by God Himself through Moses. There were intimations, crepuscular, twilight. But, but, but, thank God! the Gospel of Jesus Christ brought life and immortality to light, rescued it, brought it to light." Then the greatest reasoner this country has produced caused a sacred hush to fall upon his dying chamber; and in a loud, firm voice, he repeated the whole of the Lord's Prayer, closing with these words, "Peace on earth, and good-will to men: that is the happiness, the essence,—good-will to men." Another authority, that

---

[1] The fifty-fourth lecture in the Boston Monday Lectureship, delivered in Tremont Temple.

of his own secretary, says, that, in the last week of his life, this man, whose career you know, often repeated the whole hymn, of which the first stanza is,—

> "Show pity, Lord ; O Lord, forgive !
> Let a repenting rebel live.
> Are not Thy mercies large and free ?
> May not a sinner trust in Thee."

Webster knew his own need of these petitions. I am not here to say that he lived a Christian life. I raise this morning, when Webster is before the nation, the question, whether there is any evidence that he died repentant. I hope there is. Not many years ago I sat, on a howling winter night, at the fireside of John Taylor in gnarled New Hampshire; and he said to me, " Webster always attended the communion-service when he was at Elms Farm. Till his death he was a member in good standing with the Salisbury church, with which he united when a young man."—" But," said I, " was that church strong enough to discipline a statesman ?"—" If Webster had shown," John Taylor replied, " anything of intemperance, or other evil ways, in New Hampshire, he would have been disciplined by that church. What he did in Washington, I know not. Here, among those who knew him best, he was always ready to kneel at the family altar. There was one hymn that we always used to like to sing together," said John Taylor, with his immense bass voice, and wholly unconscious of the expression he was making of his own massiveness. "We liked to sing together ' Old Hundred :' it seemed to fit us." The venerable Judge Nesmith, whose guest I have sometimes been at Franklin, has told me things almost too sacred to be repeated here, concerning Webster's religious thoughtfulness in his last years. " Were they the last words I have to utter," said John Taylor to me, "I should say Webster died a Christian ; " and just this testimony has been given me by the profound judge, Nesmith, who stands highest among all authorities concerning Webster's life in his native haunts. Your Robert C. Winthrop, at New York on Saturday, said he had knelt with Webster at the table of our Lord, and witnessed the fervour and tenderness of his devotions.

But, gentlemen, a death-bed repentance is never to be encouraged before the time, or discouraged at the time. What I wish to insist upon, face to face with all the small philosophies of our time on both sides of the Atlantic, is the record of Webster's last speech, revised by himself. These sentences which Curtis caught are the last unrevised speech. But on Sabbath evening, Oct. 10th, the last formal speech was written, and on Oct. 15th, was revised and signed by Webster's own hand. These, his last revised words, stand upon the marble of the tombstone at Marshfield. Plymouth Rock looks on them ; and they look on Plymouth Rock. This is the record Webster left as his last word to men in all ages ; and ought it not to be copied in marble in some spot more conspicuous than that brown Marshfield shore ?

" Philosophical argument, especially that drawn from the vastness of the universe as compared with the apparent insignificance of this globe, has often shaken my reason for the faith that is in me ; but my heart has assured and re-assured me that the Gospel of Jesus Christ must be a divine reality. The Sermon on the Mount cannot be a merely human production. This belief enters into the very depth of my conscience. The whole history of man proves it."[1]

At twenty-three minutes of three o'clock on the Sunday morning following that Saturday which was illumined by the serious words on immortality, Webster passed into the Unseen Holy into which all men haste. Boston, since 1852, has been wringing her hands in secret, and saying not unfrequently, as the plain man said at the tomb in Marshfield, " Daniel Webster, without you the world seems lonesome." Are we sure that we are without him ? When Rufus Choate took ship for that port where he died, some friend said, " You will be here a year hence."—" Sir," said your great lawyer, " I shall be here a hundred years hence. and a thousand years hence."

---

[1] Curtis's Life of Webster, vol. ii. p. 684.

THE LECTURE.

If death does not end all, what does or can? If we can demonstrate by a purely physiological argument, as Draper, Lionel Beale, and Hermann Lotze, say we can, that the soul is an agent as external to the cerebral mechanism as light is to the eye, or sound to the ear, we have taken the Malakoff and Redan of materialism; and then the question is, whether we can get on in Russia. A small critic may ask how the immortality of the soul is proved by showing its externality and its independence in its relations to the physical organism. The immortality is not directly proved by the proof of the externality and the independence; but it is indirectly made probable. If you take Island No. 10 and New Orleans, you can sail from St. Louis to the Gulf, and thence to any coast you please. If, as the highest philosophy of Germany, Scotland, England, and America asserts, our nervous mechanism is wholly inert in itself, and as plainly requires an external agent to set it in motion as any musical instrument does, then the dissolution of the brain is no more proof of the dissolution of the soul than the dissolution of your organ is proof of the dissolution of the musician who plays it, but who has Gyges' ring on his finger, and is invisible. It has, in all ages, been the pretence of materialists, that the relation of the soul to the body is that of harmony to the harp, and not of the harper to the harp, or of the rower to a boat. But show me by physiological argument that the soul is an agent external to the nervous mechanism, and you have proved that the relation of the soul to the body is that of a harper to a harp, or of a rower to a boat; and, in showing that, you have removed, I affirm, not only a great, but the greatest obstacle to the belief in immortality. Unless there is evidence to the contrary, as there is not, we must believe in the persistency of that spiritual force which we call the soul; and this we must do in the name of the scientific principle of the persistence of force, itself the most vaunted of all modern points in science.

Allow me, gentlemen, to untwist a little the famous Ariadne clew, which we follow here in all our investigation; namely, that every change must have an adequate cause. In that one principle lie capsulate a great number of axioms which are at the base of all kinds of research, theological, physiological, political, or historical.

Lest you should suspect me of theological bias in untwisting the strands of this clew, take that interpretation of it which the great physiologist, Wundt, whom I have often quoted, adopts in his work on "The Physical Axioms in Relation to the Principle of Causality," a book published at Erlangen in 1866. Professor Wundt says that the principle that every change must have an adequate cause, contains in it these six axioms:—

1. All causes in Nature are causes of motion.
2. Every cause of motion is external to the object moved.
3. All causes of motion work in the direction of the straight line

uniting the point from which the force departs with the point upon which its operation is directed.
4. The effect of every cause persists.
5. Every effect is accompanied by an equal counter-effect
6. Every effect is equivalent to its cause.[1]

Will you remember, my friends, that the definition of force is this, *That which is expended in producing or resisting motion?* That is Meyer's definition; and Meyer, if he had never given any other proof of genius than this one phrase, would deserve to be called a man of great powers. But go behind even this definition, and, for the sake of clear ideas, ask what is expended in producing or resisting motion. Surely the only thing we can think of as being expended thus is *pressure.* What produces pressure? Your Carpenters, your Agassizes, and your Herschels, your Newtons, your Sir William Hamiltons, your Danas, as well as your Richters and Carlyles and Lotzes, all hold that behind the pressures which produce the motions of the universe is—WILL! MOTIONS, PRESSURES, WILL—is the universe transfigured? This is not declamation, however, but established philosophy of the latest date. Whoever will look into the last chapters of Dr. Carpenter's "Mental Physiology," or at the last sentence of Mr. Grove's famous "Essay on Correlation of Forces," or into Professor Agassiz' "Essay on Classification," or into Sir John Herschel's "Astronomy," or Dana's "Geology," or Professor Pierce's great work on "The Mathematics of Astronomy," will find the doctrine unhesitatingly maintained, that *force is always and everywhere of spiritual origin.* When I was in Harvard University, I went one day into a bookstore, and turned over a great quarto on "The Mathematics of Astronomy" by Professor Pierce; and I came upon a chapter entitled "The Spiritual Origin of Force." I looked into it; and, welling up out of that stern granite of mathematics, I found the Castalian spring of crystalline water, where the Goethes, and Herschels, and Carpenters, and Agassizes, and Lotzes, and Danas, and Richters, and Carlyles have drunk so long. In the transfiguring scientific certainty that all force originates in Will, I found that better than Delphic spring, one deep draught of which gives a new vision to the eyes, and makes the whole universe a burning bush, of which Orion and the Seven Stars are only a lowermost leaf, but every fibre of which, near and far, burns with a fire that cannot be touched, and every dustiest path before which is ground so holy, that on it we must take off our shoes, however proud of intellect we may be. Take now, the doctrine, that wherever we find heat, light, electricity, we infer motions of the ultimate particles of matter as the cause; and that, wherever we find motions, we infer pressures as the cause; and that, wherever we find pressures, we infer WILL as the cause,—and you have the point of view of these six axioms, which, by the way, are not the words of any small philosopher, nor of a theologian, nor even of an ethical teacher, but of a man simply of the microscope and scalpel, adhering

---

[1] Professor Wilhelm Wundt, On the Physical Axioms in Relation to the Principle of Causality. See, also, Ueberweg's History of Philosophy, passages on Wundt.

in all the labyrinth of modern physiological investigation, only to the idea of sanity, that every change must have an adequate cause. You say that this is poetry, and so it is ; but it is also cold, exact science. You say this is not Harvard University. Are you sure ? Yonder on the banks of the Charles sits the most philosophical poet of our generation, yes, the most philosophical on either side of the Atlantic ; and, in the name of Harvard University, James Russell Lowell might rise and sing what he sang in his own name only yesterday :—

> " God of our fathers, Thou who wast,
> Art, and shalt be, *when the eye-wise who flout
> Thy secret presence* shall be lost
> In the great light that dazzles them to doubt,
> *We who believe Life's bases rest
> Beyond the probe of chemic test*,
> Still, like our fathers, feel Thee near."
> LOWELL, *Atlantic Monthly, December* 1876.

I hold in my hand an important and enticing book, eagerly waited for by me for one, and off which the spray of the gray sea has hardly yet been shaken. It is a volume on ".The Functions of the Brain," issued only last month by Dr. David Ferrier, fellow of the Royal Society, and professor of forensic medicine in King's College, London ; and it will need no recommendation to gentlemen of the medical profession, who are permitted to know something of living tissues, and to form and express opinions after study as to the great controverted theories in biology, as no layman in science is—except the editor of the *Nation*. Professor Ferrier is a follower of two great German investigators, Fritsch and Hitzig ; and his work and theirs undoubtedly constitute not only the freshest, but the most important of all recent contributions to the knowledge of the nervous system.

Let me now, in the name of the latest research, put before you, step by step, an argument exclusively physiological, and leading up, as that of last Monday did, along this line of Wundt's wholly tremorless axioms to the conclusion that the soul is external to the nervous mechanism, which it sets in motion.

1. Fritsch and Hitzig and Dr. Ferrier have proved that certain of the convolutions of the brain of a living animal may be electrically stimulated so as to produce in the animal various physical actions.

2. The stimulation of different parts of the brain produces different results, which can be foretold by the experimenter.

3. The doctrine of the localisation of functions in the brain is now, therefore, practically beyond dispute.

I am aware that two great physiological parties,—the localisers and the anti-localisers—occupy the field of recent investigation concerning the brain. But, if we have Brown-Sequard, Hermann, Foster, and Dupuis among the anti-localisers, we have among the localisers the now preponderating names of Charcot, Fritsch, Hitzig, Ranke, Carpenter, Ferrier, Draper, and Dalton.

When you give a rabbit chloroform, and then remove a portion of its skull, the animal suffers no pain, and consequently does not fall into such contortions as to cause the act of taking away parts of the skull to injure the delicate texture of the brain. We have succeeded at last in uncovering the living, palpitating, cerebal tissues, without disturbing their delicate machinery; and we have done this by the use of chloroform, not known in the world as an anæsthetic until a few years ago. Using electrical currents that are just distinguishable by the tip of the human tongue, and employing blunted electrodes that will not scarify the nervous webs we touch, we may stimulate the exposed brain of a living animal, and ascertain that the stimulus on different parts produces different motions. We may accurately foretell these motions, after having had a sufficient experience in such kinds of experiments. One particular part of the brain, for instance, will, if stimulated, produce the attitude of resistance in the animal; and another part, if stimulated, will cause the attitude of fear. In short, a large portion of the brain has now been investigated in this way so thoroughly, that we may affirm that it is a keyboard on which electricity may play. This effect of galvanic currents on the automatic nervous mechanism is peculiarly evident on the lower or automatic nerve-arcs. You stimulate a centrifugal automatic nerve [referring to the blackboard], and you will produce motion in the muscle attached to the correlated centrifugal fibre.

Is there any proof at all that the whole brain is a keyboard that can thus be played upon by electrical stimulation?

A portion of it more closely connected with the spinal cord than the rest is a keyboard; but does the law of the automatic portion extend to the whole mass of the brain? The nervous mechanism is divided into the influential and automatic arcs. Does this fundamental distinction hold good under the searching test of electrical stimulation?

4. It is agreed that the frontal lobes are the seat of intellect.

5. *But electrical stimulation of these highest parts of the influential nervous mechanism produces no motion.*

If there are produced in this portion of the influential arcs by electricity such tremors as cause muscular motion when produced by electricity in the automatic arcs, no motion follows in the muscles. This is a fact of vast significance; but there is another of even higher import.

6. If one hemisphere of the brain be removed, paralysis of the powers of motion and sensation follows in one-half the body.

7. *But, even when one hemisphere of the brain is removed, all the mental operations may yet be fully performed.*[1]

8. These results of electrical stimulation and of cerebral injury, being opposite in the two cases, prove that physiological causes such as are concerned in the *automatic* nervous mechanism are not to be found in operation in the *influential* nervous mechanism as it is represented by the anterior lobes of the brain.

9. The distinction between automatic and influential is made broader, therefore, by the latest scientific research.

Let us examine a little leisurely the bearing of these propositions

---

[1] Ferrier, Functions of the Brain, p. 257.

upon the great biological distinction between the automatic and the influential portions in the nervous system. The important point to be noticed [illustrating by diagrams] is, that you may stimulate with electricity an influential arc here, and not produce any motion yonder. On the contrary, touch the corresponding portion of an automatic arc, and you move this muscular fibre. Although this mechanism is called automatic, remember that it was made so by the bioplasts that wove it, and that a contractile quality was given to this muscular fibre by the bioplast that wove both it and this nerve, and tied the two together. Apply your electrode to the automatic arc, and you produce contraction; but apply your electrode to the influential arc, and you produce no contraction. There is, therefore, a difference between the structure of an influential arc and that of an automatic arc. We prove this tangibly when we try point after point of the brain and of the great nervous centres connecting it with the spinal cord, and find that the lower powers of the nervous mechanism are reflex and automatic, but that these higher frontal lobes are ocularly demonstrable not to be of that sort. When we apply to them the electrical test which produces motion elsewhere, no motion whatever is produced.

If you take away one hemisphere of the brain, what is the effect? One-half the body is paralysed. The sensation and the motion which belong to the side of the body opposite to the removed hemisphere are gone. But your mental powers continue, and exhibit in completeness all their activities. Dr. Ferrier himself is authority for the astounding fact that the action of the mind is not so bound up even with these influential arcs, that it cannot show the whole army of its powers when you take away one whole hemisphere of the brain. If that can be proved, gentlemen, it has been proved tolerably well, I should say, that there is a difference between the influential and the automatic arcs, or that between the two things there is as broad a contrast as between the two scientific names. Just that has been proved beyond dispute. It is admitted by the latest science that you can take away one hemisphere of the brain, and have complete mental action yet remaining, although you cannot take away one hemisphere without paralysing one-half of the body. If I show this, I prove that there is a distinction of great breadth and significance between the influential and the automatic arcs.

"The physiological activity of the brain," says Professor Ferrier in a most suggestive passage, "is not altogether co-extensive with its psychological functions. The brain as an organ of motion and sensation, or presentative consciousness, is a single organ composed of two halves: the brain as an organ of ideation, or re-presentative consciousness, is a dual organ, each hemisphere complete in itself. *When one hemisphere is removed or destroyed by disease, motion and sensation are abolished unilaterally,*"—that is upon the opposite side, —"*but mental operations are still capable of being carried on in their completeness through the agency of the one hemisphere.*" The individual who is paralysed as to sensation and motion by disease of the opposite side of the brain (say the right) is not paralysed mentally; for he can still feel and will and think, and intelligently comprehend with the one hemisphere. If these functions are not carried on with the same vigour as before, *they at least do not appear to suffer in respect of completeness.*" [1]

A great fact this, even when standing alone ; but add to it the test of your subtle electrical stimulus, and you find that all that is implied in the distinction between influential and automatic is borne out by these two colossal circumstances,—that stimulus on the influential arcs will produce no motion, but that it does produce complex motion if applied to the automatic arcs ; and that half of the brain may be taken away, paralysing the half of your body, while the mind continues all its operations.

10. Physiological causes do not act where they do not exist.

11. The action of the influential nervous mechanism is not, therefore, *originated* by the physical causes operating in the automatic nervous mechanism.

12. But the inertness of the mechanism in itself demonstrates that it must be set in motion by an external agent.

13. That agent must be either matter or mind.

14. It is demonstrated that the action of the bioplasts in weaving the brain, and that of the frontal lobes after they are woven, cannot originate in matter.

15. It originates, therefore, in an external immaterial agent.

16. This, which is, in part, immediately known to consciousness, is life and the soul.

17. Modern microscopical research, therefore, proves that the soul is an agent external to the nervous mechanism which it sets in motion.

18. This being proved, it is demonstrated that the relation of the soul to the body is that of the rower to a boat, or of an invisible musician to a musical instrument.

19. But it has been admitted for ages by materialists themselves, that, if this is proved, then death does not end all.

Therefore, in the present state of knowledge, the case stands thus :

20. If death does not end all, what does or can ?

" Electrical irritation of the antero-frontal lobes," says Dr. Ferrier, " causes no motor manifestations,—a fact, which, though a negative one, is consistent with the view, that, though not actually motor, they are inhibitory motor, and expend their energy in inducing internal changes in the centres of actual motor execution. . . . The development of the frontal lobes is greatest in men with the highest intellectual powers ; and, taking one man with another, the greatest intellectual power is characteristic of the one with the greatest frontal development. The phrenologists have, I think, good grounds for localising the reflective faculties in the frontal regions of the brain ; and there is nothing inherently improbable in the view that frontal development in special regions may be indicative of power of concentration of thought and intellectual capacity in special directions." [2]

In this assertion, that a four-banked organ has more musical power than one with a single bank, Ferrier is not falling into materialism ; nor is he adopting the whole phrenological map, of most portions of

---

[1] Ferrier's Functions of the Brain, p. 257, § 89.   [2] Ibid., pp. 287, 288.

which he speaks with no respect. His belief is, that a new and better map will be made some day by infinite painstaking. He asserts simply that the keys on which the anthems of intellect are played are in the frontal portion of the brain, and that this anthem is at its best when the rows of keys are the most numerous, on which our invisible musician with Gyges' ring plays.

What of the immortality of instinct? A great distinction exists between those organisms that are mere automata, or have life, but no free-wills or consciences, and the higher animals, which have both the automatic and the influential nervous mechanism. The plant and the automaton have life, but not souls in the full sense of the word. But do not facts require us to hold that the immaterial part in animals having higher than automatic endowments is external to the nervous mechanism in them as well as in man? What are we to say if we find that straightforwardness may lead us to the conclusion that Agassiz was not unjustifiable when he affirmed, in the name of science, that instinct may be immortal, and when he expressed, in his own name, the ardent hope that it might be?

Go to Agassiz' grave in mount Auburn yonder, and, at the side of the Swiss boulder which marks the spot, stand alone and read these words of his, and meanwhile send your thoughts onward into the eternities and immensities, whither, no doubt, he sent his, when he wrote in the face of the world this majestic inquiry. These are the closing sentences of one of the most remarkable passages in perhaps the most remarkable of his works,—his "Essay on Classification:" "Most of the arguments of philosophy in favour of the immortality of man apply equally to the permanency of the immaterial principle in other living beings. May I not add that a future life in which man should be deprived of that great source of enjoyment, and intellectual and moral improvement, which result from the contemplation of the harmonies of an organic world, would involve a lamentable loss? and may we not look to a spiritual concert of the combined worlds and all their inhabitants in presence of their Creator, as the highest conception of paradise?"[1]

> " It was *seventy* years ago,
>   In the pleasant month of May
> In the beautiful Pays de Vaud,
>   A child in his cradle lay ;
> And Nature, the old nurse, took
>   The child upon her knee.
> 'Come, wander with me,' she said,
>   'Into regions yet untrod,
> And read what is still unread
>   In the manuscripts of God.'
> And whenever the way seemed long
>   Or his heart began to fail,
> She would sing a more wonderful song,
>   Or tell a more marvellous tale."
>   LONGFELLOW, *On the Fiftieth Birthday of Agassiz.*

---

[1] Louis Agassiz, Contributions to the Nat. Hist. of the U. S., vol. i. p. 66 ; Essay on Classification, close of part i. chap. 1, sect. xvii.

What sings she now to this great soul which has passed into that paradise of which his worthiest conception was, that it should be a concert of the combined worlds? One cannot but recollect in the sublimity of this passage that this man was born in sight of the Alps. Of French descent, of Swiss birth, of German education, of American activity, Agassiz is now of the house not made with hands; and so large was he that, even when in the flesh, he appeared by forecast to be a citizen, not of America, or of Europe, but of the supreme theocracy, in whose presence he hoped to see a concert of the combined worlds and all their inhabitants.

Richter used to say that the interstellar spaces are the homes of souls.

Tennyson sings most subtly his trust:—

> "That nothing walks with aimless feet;
> That not one life shall be destroyed,
> Or cast as rubbish to the void,
> When God hath made the pile complete.
>
> That not a worm is cloven in vain;
> That not a moth with vain desire
> Is shrivelled in a fruitless fire."
>
> IN MEMORIAM, liii.

Is it not worth while for us, standing here at Agassiz' tomb, with Richter on our right, and Tennyson on our left, to pause a moment, and on their wings, so much stronger than ours, to look abroad a little into this highest conception of paradise? A concert of combined worlds! The Seven Stars have their planets; Orion in this infinite azure is attended by his retinue of worlds; the lightest feather of the Swan which flies through the Milky Way represents uncounted galaxies; in the north, Ursa Major guards realms of life so broad, that thought faints in passing across but one of the eyelashes of the eternal constellation as it paces about the pole unwearied; Aquarius, Boötes, Sagittarius, Hercules, each holds in his far-spread palm of sidereal fire innumerable inhabitants. What if Agassiz and Richter and Cuvier and Milton and Shakspeare, and that host which no man can number, are studying at this moment a concert of all the life of Orion and the Seven Stars, Ursa Major, and the rest, and of that forgotten speck which we, on this lonely shore of existence, call earth? The loftiest exhibition of organic life of all kinds from all worlds, and in the presence of their Creator! Would it not be an immeasurable loss to be without this concert of combined worlds? Would it not be a diminution of supreme bliss not to have union with God through these, the most majestic of His works below ourselves? Shall we, too, not hope that this highest conception of paradise may be the true one? Richter would say, if he stood here, that he hopes it may be. Tennyson says, as he stands here, that he wishes it may be. Must not we, remembering the long line of acute souls who have believed in the possibility that instinct is immortal, say, that, if it be so, it is best that it should be so? Whether it is so or not, I care not to assert: what I do affirm is, that the argument for immortality, by striking against the possibility that instinct may be immortal, is not wrecked, but glorified.

When we close our short careers, some questions that we debate as matters of high philosophy will be personal to you and to me. As we lie where Webster lay, face to face with eternity, and its breath on our cheeks, there will come to us, as it cannot come now, the query whether the relation of our souls to our bodies is that of harmony to a harp, or of the harper to the harp. The time is not distant when it will be worth something to us to remember that they who walk late on the deck of the Santa Maria have seen a light rise and fall ahead of us. The externality and independence of the soul in relation to the body are known now under the microscope and scalpel better than ever before in the history of our race. Exact science, in the name of the law of causation, breathes already through her iron lips a whisper, to which, as it grows louder, the blood of the ages will leap with new inspiration. Before that iron whisper, all objections to immortality are shattered. If, in the name of physiology, we remove all objections, you will hear your Webster, when he comes to you, and says that a Teacher attested by the ages as sent with a supreme Divine mission brought life and immortality to light. There is no darkness that can quench the illumination which now rises on the world. No ascending fog from the shallows of materialism can put out the sun of axiomatic truth. Ay, my friends, in the oozy depths of the pools where the reptiles lie among the reeds in the marshes of materialism, there arises a vapour which, as it ascends higher, that sun will irradiate, will stream through with his slant javelins of scientific clearness, until this very matter which we have dreaded to investigate shall take on all the glories of the morning, and become, by reflected light, the bridal couch of a new Day, in a future civilisation.

X.

DOES DEATH END ALL? BAIN'S MATERIALISM.[1]

"Wem die heiligen Todten gleichgültig sind, dem werden es die Lebendigen auch."—
JEAN PAUL RICHTER, *Titan*, cycle 47.

> "Five hundred doors
> And forty more
> Methinks are in Valhalla.
> Eight hundred heroes through each door
> Shall issue forth.
> . . . .
> All men of worth
> Shall there abide.
> . . . .
> The ash Igdrasil
> Is the first of trees."
> THE PROSE EDDA.

PRELUDE ON CURRENT EVENTS.

CHARLES DICKENS, toward the close of his "American Notes," says, that, when in the United States on his first visit, he was often forced by sheer amazement to ask why dishonesty, conjoined with high intellectual capacity, received so much reverence from Americans. "Is it not a very disgraceful circumstance," Dickens would inquire, "that such a man as So-and-so should be acquiring a large property by the most infamous and odious means, and, notwithstanding all the crimes of which he has been guilty, should be tolerated and sheltered by your citizens? He is a public nuisance, is he not?"—"Yes, sir."—"A convicted liar?"—"Yes, sir."—"He has been kicked and cuffed and caned?"—"Yes, sir." —"And he is utterly dishonourable, debased, and profligate?"—"Yes, sir."— "In the name of wonder, then, what is his merit?"—"Well, sir, he is a smart man." Dickens says he held this dialogue a hundred times.[2] In Dickens' name I once told this anecdote to a learned German, who replied in the spirit of the renowned German candour, and in his own name, bringing his hand down upon the table with an emphasis that made the glasses ring, "That word 'smart' will break America's neck yet, unless you break the word's neck."

Every gentleman's political sympathies I wish to treat always with as much respect as I treat my own; but as to my own I say, Perish my political party, if it succeeds by fraud!

---

[1] The fifty-fifth lecture in the Boston Monday Lectureship, delivered in Tremont Temple. [2] American Notes, chap. xviii.

We are suddenly entering, in our hundredth year, upon an as yet almost unnoticed, but subtly suggestive exhibition of one great weakness in our political system, namely, that, in close elections, gigantic political spoils tempt to gigantic political frauds. In presence of Centennial guests we are now in the midst of a war of affidavits; and it appears that the combatants are about equally able. It is no empty sign of our times that contestants for political primacy, in a territory greater than Cæsar ever ruled over cannot satisfy each other that each means to be fair. The far-seeing, fateful Muse of history, holding her pen in readiness to record what is yet to be in America, and looking on the present and coming size and fatness of party political spoils in the United States, whispers to our people anxiously the words of Shakspeare's Coriolanus :—

"My soul aches
To know, when two authorities are up,
Neither supreme, how soon confusion
May enter 'twixt the gap of both, and take
The one by the other."

There are now eighty thousand minor offices filled by party patronage in the United States. While the principle, that to political victors belong political spoils, governs our politics, eighty thousand men will be turned out of office, and eighty thousand put in, with every change of the national administration. You know that Washington turned out but eight men, Adams only four, Jefferson thirty-nine, but not one of them for political reasons, Madison nine, Munroe five, and the younger Adams only two, but Jackson six hundred and ninety. Our population, as a whole, is doubling every thirty years. Soon we shall have two hundred thousand or three hundred thousand to be turned out or put in whenever a President is elected. *Will the republic bear that strain?* You will not, you say, vote for Washington's and Jefferson's rule,—to appoint the able, promote the worthy, and never remove the worthy for merely partisan reasons. You fear that there might grow up, under such a practice, an aristocracy of office-holders. It does not seem to occur to the astute opponents of civil-service reform that such an aristocracy, as it would not be turned out or put in by party patronage, and not be changed with the administrations, would serve both political parties, and so be no aristocracy at all.

Let the nation adhere for a century longer to Jackson's accursed principle, that to political victors belong all political spoils, and what must be the effect? What if closely contested national elections occur? The spoils of party patronage are already becoming so great in the United States as to constitute, with large and often controlling portions of both political parties, wholly irresistible temptations to fraud. But the spoils grow vaster and fatter with fearful speed. *Only civil-service reform can remove this enormous coming mischief.* It can do so only by taking patronage from party, and giving it to the people. *Gigantic party political spoils, gigantic party political frauds,—these are cause and effect.* They imperil the peace of the republic. They must do so more and more as our population grows. *Ultimately in America there will be either civil-service reform or civil war.*

THE LECTURE.

Plato represents Socrates as saying that he had looked at many authorities, and, among others, at the nature of things, but dared not look long at the latter for fear his eyes would be dazzled.[1] It is the radiance of the nature of things, or axiomatic, self-evident truth, which must frighten back to Chaos the vampire Doubt. On some sickly veins of our moaning and sceptical age that vampire broods as a nightmare; but no nightmare can bear the noon. Mrs.

---

[1] Phædon.

Browning sang plaintively in the name of poetry, and her antipodes, Ernst Häckel, affirms aggressively in the name of science, that,

"A wider metaphysics would not harm our physics."

Two thousand years ago, Aristotle, with a measureless plaintiveness and gladness, wrote what the history of all discussion has since confirmed, that they who forsake the nature of things, or axiomatic first truths, will not and cannot find anything surer on which to build. Let us bring all those who are halt and lame and blind with doubt, or mental unrest, into the sunlight of axioms. Let us cheer ourselves in the vivifying radiance of the noon of the self-evident truths. The questions which the progress of science raises the progress of science will answer. It will do so, not to the detriment, but to the coronation, of religious science. Twenty centuries before the modern forms of physical science were born, religious science made, as she yet makes, the dateless and eternal noon of axioms her soul.

I find no form of materialism, old or new, that can look into the authority which dazzled Socrates, and retain steadfastness of gaze.

What is the newest form of materialism? That of Professors Bain and Tyndall, and that which is adopted, in a large degree, by Huxley and Spencer, and, almost without qualification, by Häckel. You know that St. George Mivart calls Huxley Häckel's *Alter Ego* "Contemporary Evolution." No man doubts that Häckel, in spite of his protestations, is a materialist, or one who believes that there is but a single substance in the universe, namely, matter. "The will is never free" is Häckel's constant teaching; and to his amazingly narrow philosophy, which Germany discards, "God is necessity" only, and has "no freedom of choice." Huxley quietly holds many of Häckel's philosophical opinions, but expresses them with far less boldness on their offensive side than Häckel does. When it is asserted that Herbert Spencer's positions are not of materialistic tendency, let a competent witness be called, say Thomas Rawson Birks, professor of moral philosophy in Cambridge University, England. This trained and indorsed scholar has just sent to us across the sea a work of beautiful clearness and candour, entitled "*Modern Physical Fatalism*, and the Doctrine of Evolution, including an Examination of Mr. H. Spencer's First Principles." The "Fatalistic Philosophy and Doctrine of Evolution, *as unfolded by Spencer*," he regards as "radically unsound, full of logical inconsistency and contradiction, flatly opposed to the fundamental doctrines of Christianity, and even to the very existence of moral science."[1] You must not allow yourself to think that the highest philosophical authority in Cambridge in England, and the highest in Cambridge in America, are really of two opinions as to any philosophy that is predominantly Spencerian. Is it maintained that Huxley is not a materialist in any sense, because he has said that he is not in some senses of that word of many meanings? What are his definitions? Who is it that teaches in so

---

[1] Preface, Sept. 28, 1876.

many words, in his latest and most deliberate utterance,[1] that "a mass of living protoplasm is simply a molecular machine, the *total results of the working of which, or its vital phenomena, depend, on the one hand, on its construction, and, on the other, upon the energy supplied to it* ; and *to speak of vitality as anything but the name of a series of operations is as if one should talk of the horology of a clock*" ? If that is not materialism, what is ? How much more space does that definition leave for freedom of the will and moral responsibility and immortality than is left by Häckel's more outspoken but not more sweeping phrases ? That sentence contains both Huxley's and Spencer's central position. But every redoubt in the camp which defends the mechanical theory in biological science is riddled and ploughed by the artillery of Hermann Lotze and Wundt and Helmholtz, and all the best learning of Germany, to say nothing of Scotland and America. Of course, the English materialistic school must pick its phrases carefully. It often says it is not materialistic ; but it is to be tested by its definitions. Many of Huxley's phrases imply not only a fear of arousing the aversion of scholars to materialism, but also a lack of intellectual unity. Tyndall and Huxley are both freely accused in England and Germany of metaphysical incompetence. On the question whether certain schools of thought are materialistic or not, those innocent souls who cannot fasten their eyes fixedly on definitions will find all the beaten paths of modern philosophical discussion full of what politicians call dust for the eyes of the unwary.

In the sea of axiomatic truth, materialism swims with fins of lead.

1. Bain's and Tyndall's materialism, which is the latest and subtlest kind, asserts that matter is "a double-faced unity," having "two sets of properties, or two sides,—the physical and the mental ;" but is, nevertheless, "one substance," and the only substance which exists in the universe.[2]

2. If this definition is correct, it follows, that, in matter, physical and spiritual qualities must not only inhere, but *co*-inhere, in one and the same substratum. The qualities of matter and mind must be conjoined in one substance.

3. Among the fundamental qualities of matter are extension, inertia, gravity, colour, form.

4. But the qualities of mind are the antipodes of these qualities. It is absurd to speak of the extension, inertia, gravity, colour, or form of a thought, an imagination, a choice, or an emotion. When Cæsar saw Brutus stab, and muffled up his face at the foot of Pompey's statue, was his grief round, or square, or triangular ? When Newton conceived the idea that gravitation is a universal law, was that thought red, or brown, or violet ? When Lincoln by a stroke of his pen manumitted four million slaves, was his choice hexagonal, or octagonal ? Does the act of imagination in a Shakspeare weigh an ounce, or a pound ? These questions show that the terms which we apply to matter are totally inapplicable and meaningless if applied to mind.

---

[1] Huxley, Encyc. Brit., art. "Biology," 1875. [2] Bain, Mind and Body, p. 196.

5. Professor Bain himself admits that the organic and the inorganic are not so widely separated as matter and mind; and that the elements of our experience are in the last resort, not one, but *two*. "Mental and bodily states are utterly contrasted; and our mental experience, our feelings and thoughts, have *no extension*, no place, no form or outline, no mechanical division of parts, and we are incapable of attending to anything mental until we shut off the view of all that."[1]

You must not suppose that Bain is witless enough not to recognise the distinction between mind and matter as the broadest known to man. His work on "Mind and Body" I hold in my hand; and it is one number of those royal and very disappointing roads to knowledge, called "The International Scientific Series." I reverence Professor Bain. He has written some books which are thorough, and will bear, in most parts, the logical microscope. But this volume on "Mind and Body" seems to have been made to order and in haste. Nevertheless, it is the Bible of the latest English materialism; and now, out of this freshest revelation, let me read a text or two.

"EXTENSION," says Professor Bain, "is but the first of a long series of properties *all present in matter, all absent in mind.* INERTIA cannot belong to a pleasure, a pain, an idea, as experienced in the consciousness. Inertia is accompanied with GRAVITY, a peculiarly material quality. So COLOUR is a truly material property: it cannot attach to a feeling, properly so called, a pleasure or a pain. These three properties are the basis of matter; to them are superadded Form, Motion, Position, and a host of other properties expressed in terms of these, Attractions and Repulsions, Hardness and Elasticity, Cohesion, Crystallisation. Mental states and bodily states cannot be compared."[2]

These sound very much like Sir William Hamilton's phrases, but they are Bain's; and yet, turn on to the last and most emphatic paragraph of this book, and you find a proposition at which Sir William Hamilton or Hermann Lotze would only smile; namely, that there is in the universe but "one substance," which has two "sides,"—whatever that word may mean,—"a physical and a mental," and so is "a double-faced unity." "The arguments for the two substances have, we believe, now entirely lost their validity. The one substance with two sets of properties, two sides,—the physical and the mental,—a double-faced unity, would appear to comply with all the exigencies of the case."[3]

Not if the nature of things is yet as dazzling to us as it was to the eyes of Plato and Socrates and Aristotle and Liebnitz and Kant and Hamilton; not if axiomatic truth is as radiant to us as it is to Lotze and Helmholtz and Wundt and Beale and Dana; not if we are to adhere to the first of all logical laws, that, whatever stands or whatever falls, a thing cannot be and not be at the same time and in the same sense.

---

[1] Professor Alex. Bain, Mind and Body, pp. 124, 135.
[2] Ibid., pp. 125, 135. [3] Ibid., p. 196.

6. If matter is a double-faced unity, having a spiritual and physical side, there must co-inhere in one and the same substratum extension and the absence of extension, inertia and the absence of inertia, colour and the absence of colour, form and the absence of form.

7. To assert that these fundamentally antagonistic qualities of matter and mind not only inhere, but co-inhere, in one and the same substratum, is to assert that a thing can be and not be at the same time and in the same sense.

8. This limitless self-contradiction wrecks in this age, as it has wrecked in every age, the pretence that there is but one substance in the universe.

9. We know incontrovertibly that there are two sets of attributes, which, as diametrical opposites, cannot co-inhere in one substance, since extension and its absence, inertia, form, colour, and their absence, cannot possibly co-exist in one and the same thing at the same time.

10. Every attribute, however, must belong to some substance.

11. *Two irreconcilably antagonistic sets of attributes must belong to two substances.*

This proposition is as venerable as the sword Excalibur of King Arthur. With it materialism of the older forms has been defeated on many a Waterloo of philosophy; with it materialism in its newest form has no battle but that which consists in flight from its deadly edge.

12. The axiomatic knowledge we have of two such sets of attributes necessitates the conclusion that matter and mind are two substances.

13. In that inference from self-evident truth, all forms of materialism are shown to be absurd, as all forms alike assert that there is but one substance.

14. *Professor Bain's fundamental error is the confusion of "close succession" with "union."* He asserts "union" of the qualities of matter and mind in one substance with two sets of properties. He endeavours, but in vain, to show that this is not union in place ; and then says,[1] that "*the only mode of union that is not contradictory is the union of close succession in time.*" Such *succession* is not *union* in any sense that can justify the assertion that there is but *one* substance in the universe with two sets of properties.

In the last pages of this weak book, Moleschott, Vogt, and Büchner, whom Germany regards as little men, are mentioned as among the recent bright lights of materialism. Bain admits distinctly, and yet, of course, without emphasis, that "*it is not to be supposed that these writers are in the ascendant in Germany.*" His poor sketch of the history of materialism is intended to show that this system of thought may expect a successful future. That argument, however, with many others, stumbles, and falls flat over his concession that the most intellectual nation, in which philosophy is a passion with scholars, and which has given to this subject more thought than

---

[1] Professor Alex. Bain, Mind and Body, p. 137.

all other nations combined, repudiates the latest as well as the oldest materialism.

Gentlemen, I know that thus far in this address the argument is metaphysical; but, in the audience of scholars, it is not for that reason useless. Metaphysics is simply *an articulate knowledge of the necessary implications of axiomatic truths,* and is not only a very clear and exact science in itself, but the mother of all the other sciences. We must reject either self-contradiction or sanity. We must adhere to primary, self-evident truths, or fall into that ultimate form of scepticism which knows nothing except that it knows nothing, and does not know even that except upon the evidence of these very axioms or intuitions, with which it plays fast and loose. The man who does not know much is a great character in our inquiring but unphilosophical times. When you trace a mind which rejects axioms up to its last refuge of oleaginousness, or ignorance, or weakness, you can ask, "Are you sure that you know nothing with certainty?"—"Yes," he replies, "I am sure."—"But then there is one thing you know with certainty."—"No; I am sure that I know nothing surely."—"But how are you sure that you are sure?" Only on the authority of the axiomatic, self-evident truths which dazzled the eagle eyes of the Acropolis; are presupposed in all reasoning; and are imbedded not only in the human mind, but in the very nature of things. Every change must have a cause. The whole is greater than a part. Mind exists. Matter exists. A thing cannot be and not be at the same time and in the same sense. A straight line is the shortest distance between two points. These are a few of the renowned fundamental principles, first truths, axioms, intuitions, eternal tests of verity, of which metaphysics gives the list; and to conscientious consistency with these, it is the duty of religious science, which first elaborately studied axioms, to hold mercilessly all other sciences and herself.

Curiously, and yet not curiously, physiology, and metaphysics tell the same tale whenever they speak on the same points. To test one science by another is the most important, and, intellectually, the most delicious, of all arts. Let us turn now to physical, concrete facts again, and observe the coincidence of their testimony with that of the primary mental facts or axioms. In the field of modern physiological research, materialism fails through hopeless and practically measureless self-contradiction.

1. If matter is a double-faced unity, having a spiritual and physical side, and is the only substance that exists in the universe, then, in matter, spiritual and physical qualities must not only inhere, but *co-*inhere, in the same substratum.

2. It must be true of every atom of matter that it has a spiritual and a physical side.

3. In every atom, therefore, spiritual and physical qualities must be found so inseparably conjoined, that the one side cannot be conceived to be taken away without carrying the other side with it.

4. If this be the true character of matter, then the physiological activities of the atoms must be at least co-extensive with the psycho-

logical activities displayed in connection with those atoms; that is, *both the psychical and physical sides of the one substance-matter must go together, and, if the latter be removed from any grouping of atoms, the former must go with them.*

5. According to this newest materialistic definition of matter, the physiological activities of the brain must be *in this sense* co-extensive with its psychological activities.

6. But according to the experiments of Ferrier, Fritsch, and Hitzig, one whole hemisphere of the brain may be taken away, and one-half the body paralysed in consequence, and yet the mental operations remain complete.

7. "The physiological activities of the brain are not co-extensive with its psychological activities."

This is Ferrier's own language, of which he does not seem to see the philosophical importance.

8. Matter, therefore, is physiologically demonstrated not to be a *double-faced unity with inseparably conjoined spiritual and physical properties.*

9. But the psychological changes taking place in the mind must have an adequate cause.

Evolution equals involution. There cannot be in the effect what does not exist in the cause: if there could be, there would be an effect without a cause.

10. The adequate cause of the psychological changes taking place in the mind does not exist in the physiological changes going forward in the brain; for, *other things being equal, effects must vary when their causes vary; and the half of the brain may be taken away, and the mind yet perform with completeness all its operations.*

Many writers have taught that the connection of cause and effect may be tested in three ways,—either by taking away the cause, and noticing that the effect ceases; or by introducing the cause, and noticing that the effect springs up; or by making the cause vary, and noticing that the effect varies. We cannot take the moon out of the heavens, and we cannot dip the tides out of the sea; and so, in regard to the tidal motions of the ocean, we cannot apply the first two of these tests. But we can use the third; for we notice, that, when the sun and moon are in conjunction, the tides are higher than at other seasons. We observe that the tides follow the moon, and always vary according to its position. Now, this is precisely the test that I apply in reading under the law of causation the philosophical import of the latest physiological facts. We cannot take apart the body and soul, and then bring them into conjunction, noticing first the effect of their separation, and then that of their union; but we can cause the one to vary somewhat, and notice the variation, or absence of variation, in the other. We take away a hemisphere of the brain, and do not produce the variation in the mind which it is perfectly clear ought to follow if materialism is true. Bain's pretence, that the antagonistic qualities of matter and mind inseparably co-inhere in one substance-matter, is inconsistent with such a fact as Ferrier brings before the world, when he says, as all physiologists say, that you may take half a brain away, paralysing half the body, and yet leave the mental operations—memory, imagination, affection, choice, reason, perception, the whole list of faculties—complete. We vary the supposed cause, and the supposed effect does not vary ; and this is proof that it is not an effect.

It is to be expected that a small diminution of vigour in mental action may follow the taking away a hemisphere of the brain ; but in a large brain this effect is hardly perceptible. Take away half the force of the bellows of your organ yonder, and your anthem proceeding from the organ is less loud ; *but all its notes and rhythms remain. In the brain is your anthem in the bellows, or in the musician's fingers ?* Materialism is a stupid peasant that for ever stands behind the organ, and can see only the bellows, and never the musician ; and asserts, when the latter wears Gyges' ring, that he does not exist, and so would blunderingly account for the anthem by the bellows and organ alone.

11. As the adequate cause of physiological changes in the mind cannot be found in matter, it must exist outside of matter.

Hermann Lotze is for ever reiterating as the great maxim of his philosophy, " Exceptionally wide in the universe is the extent, entirely subordinate is the mission, of mechanism." This is the keynote of the deepest philosophy of Germany at this moment, that mechanism is to be found everywhere in the universe, but that it is everywhere the horse, and not the rider. " Exceptionally wide in the brain," Hermann Lotze would say, " is the extent, but wholly subordinate is the mission, of the nervous mechanism."

We must remember that this very mechanism, the known origin of which is left in such mystery by materialists, is woven by the bioplasts with a sufficient cause behind them. We must study that cause by its phenomena, as we study any other object in Nature. As many unprejudiced students as have seen Lionel Beale's preparations and exhibitions of tissues under the microscope, have, he says, hopelessly abandoned materialism.

A fascination not easily described attends the study of living movements under the microscope, as a kind of conviction there comes to you, which no diagrams convey, that life and mechanism are two things. I am properly conscious of the fact that I am no microscopist. Perhaps I had better reveal, however, that it happens that I have the opportunity to use, at any hour of the day or night, what I suppose to be by far the best microscope in Boston. It belongs to a professor, a physician, who has made histology a specialty, and who was so kind as to invite me to use his magnificent instrument. It is what the books call a one-seventy-fifth objective ; and the highest power Beale is using is only a one-fiftieth. This prince among microscopes is in Tremont Temple building now ; and it shows a white blood corpuscle nearly as large as the silver piece called a sixpence ; and even Lionel Beale's best instruments show it hardly larger than a three-cent piece. Dissections of brains are offered to my inspection frequently ; and, although I have no right as a student of religious science to do so, I seize eagerly every opportunity to study the physiological side of philosophy as one part of religious science. Let me say that only the other evening, in this very Temple, in company with experts who all believed in Lionel Beale, and not in the mechanical theory of Häckel, I saw living bioplasm pass and repass through the field of this exceptionally excellent instrument. I had read all Beale says of bioplasmic movements ; I had impressed upon myself the intricacy of the work done by the bioplasts ; I had minutely studied the best coloured plates ; and I thought I knew something of the difference between the action of life and that of merely physical force : but, when I saw bioplasm itself in movement [such as is represented here], I felt myself in the presence of an entirely new revelation of the inadequacy of materialism, with all its prate about chemical forces, to account for the weaving, I will not say of a brain, an eye, an ear, or a hand, or of nerve within nerve, and of bone beneath muscle, but of the humblest and simplest living fibre that ever a bioplast spun.

Think of the various activities of the one substance bioplasm ! The fluid that lubricates the eye is thrown off by the same matter that constructs bone. The muscle and the tendon are woven on one loom. Take that which you drink at your tables, and call milk, and what is it but smooth cell-walls thrown off by the bioplasts, and now, in their absence, sliding over each other as a beautiful fluid ? What is this instrument of three thousand strings, which we call the ear, but a mass of cell-walls woven together by bioplasts ? How are we to account for the miraculous retina and lenses of the eye ? They came from the same loom that weaves the brain. *How is such variety of effects to be accounted for with no variety of mechanism ?*

12. Outside of matter is to be found only what is not matter, that is, an immaterial cause.

13. The existence of that cause is demonstrated by the application of the axiomatic truth, that every change must have an adequate cause.

14. This same law demonstrates the externality and independence of this cause in its relations to the cerebral mechanism.

15. The relation of this immaterial agent to the body, therefore, is that of a harper to a harp, or of a rower to a boat, and not that of harmony to a harp.

16. The dissolution of the brain, therefore, no more implies the dissolution of the soul than that of a musical instrument does that of an invisible musician who plays upon it, or that of a boat does that of the rower.

17. Death, therefore, does not end all. Therefore, for the third time, by an independent line of argument purely physiological, we conclude,—

18. If death does not, what does or can ?

To outline now a third argument, let me ask you to notice in all their relations to each other this series of propositions :—

1. It is a physiological fact that every human being once breathed by a membrane, then by gills, then by lungs, and once had no

heart, and then a heart with but one cavity, and then a heart of four cavities.[1]

2. The particles of the body are continually changing.

3. In the metamorphoses of insects, not only are the particles of the body changed, but its entire plan is altered.

Will you, my friends, but picture to yourselves the change of plan which must be made when a creeping creature is transformed into a flying one? Your beautiful tropical butterfly was once a repulsive chrysalid. It had only the power of crawling. But the bioplasts wove it. Little points of transparent, structureless matter were moving in it, were throwing off cell-walls in it, and bringing these walls into the shape, now of a tendon, now of a muscle, now of a nerve, and so completing the whole marvellous plan of a crawling creature; disgusting in our first sight, a miracle at the second. But now these same bioplasts, which, according to materialism, have nothing at all behind them but chemical forces, suddenly catch a new and very brilliant idea, namely, that they will weave a flying creature. Whence comes that? Out of matter; for matter has a physical and a spiritual side. They thereupon, without any new environment, with the same sun above them, and the same earth underneath them, and the same food, begin to execute a wholly new plan, or rather to carry out one held in reserve from the first. They weave anew; there appears within, and rising out of, the creeping, odious worm, your gorgeous tropical butterfly; and *he is the same.* There is identity between that flying creature and that creeping creature. Are they two, or one? You breathed by gills once; you breathe by lungs now. Is your identity affected in the change? Your bioplasts wove you, so that once you had a heart of one cavity, and now have one of four. Are you the same? Is your identity affected through all these changes? Every few months, the flux of the particles of the living tissues carries away all the particles in the entire physical system. How do we retain identity? Matter has a physical and a spiritual side, indeed. While all the matter that composed my body has gone in the flux of growth, I am I, however. I have an ineradicable conviction that I am the same person that I was years ago; and yet, years ago, there was not in my body a particle that is now there. I have an ineradicable conviction that the butterfly is identical with the crawling worm; but the characteristics of your worm are left behind when there appears in the worm a resurrection to a new life.

What if your butterfly were in all his parts as invisible as he is in some portions of his wings; and what if, to human ken, through sight or touch, there could be no account given whatever of that creature woven out of the loathsome chrysalid? What if, out of that discarded organism, were to arise something equally glorious with the butterfly, but wholly invisible, would this change be more miraculous than the rising of that visible winged creature out of that body? I think not. *If God can lift the visible out of the chrysalid, may He*

---

[1] Draper, Physiology, p 550.

not be able to lift the invisible also? Yes; but you say that this is Christian materialism. I beg your pardon: I know what thoughts beyond the reaches of our souls rise for utterance as we face life in death. I do not assert that the soul is material; nor do the Scriptures do so, where they affirm that there is a spiritual body as there is a natural body. What that means, I need not here, in the presence of so much learning greater than mine, discuss; but I do affirm, that if God, instead of lifting a visible, were to lift an invisible, flying creature out of the worm,—insect or man!—He would perform no greater miracle than that He does now. Nothing more inconceivable would it be to lift a wholly invisible new form out of a chrysalis than one partially invisible. The change need not be greater; and He who can do the one miracle, and does it day after day before our eyes, can do the other.

4. In all the flux of the body the soul retains conscious, personal identity.

5. The unity of consciousness, and the sense of continuous personal identity, require adequate explanation.

6. Nothing can exist in an effect which did not previously exist in the cause.

7. Effects must change when causes change.

8. If conscious personal identity were an effect of the matter comprising the physical organism, it ought to exhibit as an effect the same flux which exists in its supposed cause.

9. No such flux is observed in the effect.

10. Therefore, the cause of the sense of personal identity is not to be found in the matter of the organism.

11. As only matter and mind exist in the universe, that cause must be an immaterial agent existing in connection with the physical organism.

12. That agent is known to consciousness, and is called the soul.

13. Its existence is not only known to consciousness, but is demonstrable by the law of causation, which requires that every effect must have an adequate cause.

The unity of consciousness and the permanence of personal identity are supreme German arguments against all forms of materialism.

This is the birthday of Thomas Carlyle. Eighty-four years ago, in the stern year in which Louis XVI., Marie Antoinette, and Charlotte Corday, went to the scaffold, there came into the world the first prose poet of our time, and the most lofty and vivid imagination, except Richter's, since Milton. Is it not fitting that on this day, at least, we should listen seriously to a man who has thought boldly, and with no narrow mental horizon?

"You have heard," says Carlyle, and in perfect freedom from all bias but that of genius, "St. Chrysostom's celebrated saying in reference to the Shechinah, or ark of testimony, visible revelation of God among the Hebrews: 'The true Shechinah is man.' Yes, it is even so: this is no vain phrase; it is veritably so. The essence of our being is a breath of Heaven. This body, this life of ours, these faculties, are they not all a vesture for that Unnamed? *We touch Heaven when we lay our hand on a human body.* We are the miracle of miracles. This is scientific fact. God's creation—it is the Almighty God's. Atheistic science babbles poorly of it with scientific nomenclatures, experiments, and whatnot, as if it were a poor dead thing to be bottled up in Leyden jars, and sold over counters; but the natural sense of man in all times, if he will honestly apply his sense, proclaims it to be a living thing. Ah! an unspeakable, God-like thing, toward which the best attitude for us, after never so much science, is awe, devout prostration, and humility of soul; worship, if not in words, then in silence."[1]

Who in Boston has a right to look loftily on Carlyle? Macaulay said, but let me only whisper the fact, that he did not see how Prescott, being what he was, could live in such a place as Boston. Who in any American editor's chair, or in any college in New England, is authorised to look condescendingly upon Carlyle, even on this theme, although, forsooth, he is not a microscopist?

---

[1] Carlyle, Hero Worship.

XI.
AUTOMATIC AND INFLUENTIAL NERVES.[1]

"It is certain that matter is somehow directed, controlled, and arranged, while no material forces or properties are known to be capable of discharging such functions. . . . I believe that it will be found, that the institution of the series of preparatory changes which occur previous to the development of the lasting form and structure of tissues can only be accounted for upon the supposition of the existence of a power capable of fore seeing what was about to happen, and of determining beforehand the arrangement that would be most advantageous to the living being, and able to provide beforehand for requirements that it was foreseen would arise at a future time."—LIONEL BEALE, *Protoplasm*, pp. 306, 358.

"The laws of nature do not account for their own origin."—JOHN STUART MILL, *Logic*.

PRELUDE ON CURRENT EVENTS.

IT is sometimes sneeringly affirmed that colleges teach little but the art of finding where knowledge is; and yet that is a great and difficult art. In the froth-oceans of weak books, it is a high service to point out to a hurried man, on any interesting theme, the most serviceable volumes. What are the dozen best English, and what the dozen best German books on biology? In response to many inquiries, verbal and written, let me attempt an answer to this rather formidable question. There are few or no good books on biology older than 1860. Remember that the microscope did not attain its power to furnish facts of a scientific character for the basis of research till 1838. So fast has the study of living tissues progressed, that it may be said that all the conclusions reached before 1860 either have been or will be modified. I therefore can recommend to you nothing older than 1860, except an author or two like Schleiden and Schwann, who began the investigations of living tissues, and whose works are to be examined for their interest as historical documents. On this theme, as on so many other philosophical matters, the best books are German; but take first the English in the order of their merit:—

1. Beale, Dr. Lionel S., "Protoplasm; or, Matter and Life." Third edition: London, G. & A. Churchill; Philadelphia, Lindsay & Blackiston, 1875.

The style of this work is attractive for its clearness, grace, and force, and occasionally for a keen, logical humour. It is not always that a physician has literary capacity; but Lionel Beale is a good and almost a brilliant writer. Besides, he has had a liberal training in logic and metaphysics, and seems to have grasped philosophy as a whole very fully. But the charm of his book is in the luminousness, vivacity, and power produced by his stalwart grasp of his theme as an original discoverer. No doubt he has added more to the knowledge of living tissues than any living English author within the last twenty-five

---

[1] The fifty-sixth lecture in the Boston Monday Lectureship, delivered in Tremont Temple.

years. It does not become me to state here what precautions I have taken to know that I have not been misled in seeking authorities on biology; but I have taken precautions of a most merciless sort, and continue to take them, and all my precautions end in giving me more and more confidence in Beale, as a man of candour and sense as well as of science. If you can buy the productions of but two authors on biology, purchase the works of Beale as the best that the English language offers you, and those of Frey as the best that the translated German at present affords.

2. Frey, Professor Heinrich, Zurich, "Manual of Histology," Leipzig, 1867; and "Compendium of Histology," Zurich, 1876. Translated by Dr. George R. Cutter. New York: Putnam Sons, 1876. Frey's two works are by common consent placed now at the head of German works on histology.

3. Drysdale, Dr. John, "The Protoplasmic Theory of Life." London, 1874. This work of an Edinburgh physician, and president of the Liverpool Microscopical Society in 1874, seems to stand third in order of importance. It does not adopt Beale's conclusions as to vital force; but it accepts his facts, and makes a strenuous and futile effort to reconcile them with what is called the theory of stimulus.

4. Ferrier, Dr. David, "The Functions of the Brain." London and New York, 1876. This work is indispensable to any one who does not read German books on biology.

5. Tyson, Dr. James, "History of the Cell Doctrine."

6. Carpenter, Dr. W. B., "Mental Physiology." London and New York, 1874.

7. Beale, Dr. Lionel S., "How to work with the Microscope." New edition. Philadelphia, 1877.

8. Kölliker, "Manual of Human Histology." Translated by George Bush and Professor Huxley for the Sydenham Society, 1853.

9. Huxley, Professor T. H., art. on Biology in ninth edition of "Encyclopædia Britannica."

10. Carpenter, "Human Physiology," ninth edition, 1876.

11. Draper, Professor J. W., "Human Physiology," 1856.

12. Dalton, Professor John C., "Human Physiology," edition of 1875.

Here is a list of twelve German authors :—

1. Lotze, Hermann, " Mikrokosmus," 3 vols., 1873. Lotze was born at Bautzen in 1817. He was graduated at Leipzig in 1834, in both philosophy and medicine. In 1842 he became professor of philosophy at the University of Leipzig, but since 1844 has been professor of philosophy at the University of Göttingen. His collected works are to be recommended as all bearing on biology.[1]

2. Ulrici, "Gott und die Natur." Halle, 1873. "Gott und der Mensch." Leipzig, 1874.

3. Stricker, "Handbuch der Lehre von der Geweben des Menschen und der Thiere. Leipzig, 1868.

4. Häckel, "Generelle Morphologie der Organismen," 1866.

5. Schultze, Max, "Protoplasma der Rhizopoden," 1863. Read all of Schultze's works.

6. Neumann, " Ueber d. Zusammenhang sog. Molecularen mit dem Leben des Protoplasma;" Du Bois-Reymond and Reichert's "Arch.," 1867.

7. Kölliker, "Neue Untersuchungen," &c., 1861.

8. Kühne, W., "Untersuch. über das Protoplasma," 1864.

9. Helmholtz, "Handbuch der physiol. Optik."

10. Wundt, Physiologie des Menschen.

11. Hitzig, " Untersuchungen über das Gehirn."

12. Du Bois-Reymond, Ueber die thierische Electricität.

Omitted books which scholars here may think I ought to have named, would probably appear if I were to give a list of the hundred best volumes. If you can buy but three books, have Frey's "Histology," and Beale's "Protoplasm," and Lotze's "Mikrokosmus."

---

[1] See art. on "Hermann Lotze," in July number of Mind, 1876.

THE LECTURE.

If Aristophanes were here to-day, we perhaps could give him no better entertainment than to cause a frog to utter the famous words of one of this Greek poet's plays: *Brekekekéx, koáx, koáx*.[1] We might take a brainless frog, and, by gently stroking its back, we should produce these Greek words, uttered automatically by the vocal organs of the amphibian; and, as often as we stroked the back, we should insure that result. Goltz, the German physicist, who has lately written an elaborate work on the nerve-centres of frogs,[2] says very genially that the batrachian chorus of our summer evenings is the natural proclamation of the fact that it is well with the inhabitants of the marsh as the sedges and the ooze stroke their backs under the still stars. I am not supposing our frog's brain to be removed as a whole, but so far forth only as the taking-away of what are called the cerebral hemispheres can change the mechanism of the complex nervous mass within the skull. The lower nervous centres in the spinal column and in the neck, and just above, remain in the frog. When I pinch him, thus brainless, he leaps. When I place the miraculous creature in the palm of my hand, and turn the hand, as Huxley did his in a famous public experiment, intended, but not sufficient, to puzzle the world as to the freedom of the will, the frog keeps position, and stands upon the back. I reverse the motion, and he keeps his place, and stands upon the palm. This is not an effect of will on his part, but of the life which stands behind that marvellous automatic mechanism which his bioplasts have woven. I put him in his native pool, and he swims the instant he touches the water. On reaching the shore, however, he at once becomes quiet. He sits there hours and days; and, if he is not again touched by some external force of such a kind as to irritate his automatic nerves, he will seek no food, and will continue quiet until he becomes a mummy. All this looks as if the frog were an automaton; and so, indeed, he is when the hemispheres of the brain are taken away. But, when these hemispheres are present, the frog seeks food; he does not sit in one spot; his automatic croak he represses when a stone is thrown among his watery bowers of grass and reeds; he has multitudinous playful ways; he possesses, in short, the power of self-direction. All this he loses with the removal of the hemispheres. The animal that has lost these, however great its remaining automatic power may be, will not seek food, and, unless artificially fed, always perishes of starvation. There appears to be nothing like choice or volition left in the frog after the cerebral hemispheres are ablated.

Take a fish, and remove its cerebral hemispheres, and you will find that the same great contrast between automatic and influential nervous action appears. The fish swims with perfect equilibration. The stroke of the fins and tail retains its amazing precision. But the mutilated swimming creature does not stop, as other fishes pause,

---

[1] Aristophanes' The Frogs.   [2] *Functionen der Nervencentren des Frosches*, 1869.

to nibble at food here and there. It does not loiter, as its companions do, in shaded aqueous couches. It flashes not up thence, as they do, to catch the unwary insect in the evening or morning dusk. The brainless fish has no capacity to play in spheral rhythm with its mates and with the waves. It keeps on in a straightforward course, unless turned aside by some obstacle; and does not pause until nervous or muscular exhaustion necessitates rest. That fish, too, will perish of starvation unless artificially fed. It has no tendency to seek food; its volitional power is lost. In this case of the fish, a very different law would seem to be exhibited from that which appears in the case of the frog; and yet the two cases are to be explained by precisely the same contrast between the automatic and the influential nervous arcs. The fish has a constant stimulation of the automatic nerves. The water produces reflex movements; and these, so wisely did the bioplasts of the fish weave the creature, constitute the complex act of swimming. Your frog sits still because no stimulus is applied to the automatic nerves; and your fish swims because a prolonged excitation of those nerves is produced by the water. But, to show that the case of the frog and that of the fish are parallel, put the frog into the water, and he will swim in it as long as it floats his body. He is an amphibious animal, and will get out upon the land if he can; and this is the only difference in the case.

Let us remove from a pigeon the central hemispheres, and we shall find that the poor bird, when we wave a fiery brand in a circle before its eyes, will follow the motions of the light with its head. If a fly pauses on its crest, it will shake off the intruder. Placed on its back the bird will regain its feet. If it walks along your table, and comes to the edge, it will lift its wings the moment this action is necessary to balance its form. So mysteriously have its bioplasts woven this flying creature, that, when the pigeon thus brainless is cast out upon the free air, it moves there with its accustomed royalty, as if in its home. But when left at rest it makes no spontaneous movements. This brainless bird, like the brainless frog or fish, unless stimulated by some outward touch, remains for ever quiet, never seeks food, and will become a mummy. It has apparently no power of originating muscular action. It possesses the lower nervous arcs; but you have taken away the upper, and in doing this you have taken away its power of originating movements.

Removal of the hemispheres from a rabbit leaves the animal for a while prostrate; but, after a varying interval, it exhibits power to maintain its equilibrium on its legs in an unsteady manner. A loud sound causes its silken, sensitive ears to twitch, its quivering, aspen-leaf body to start. Its flight, once begun under proper stimulation, is headlong, bungling, and blind. If left to itself, it will seek no food, remain fixed and immovable on the same spot, and, unless artificially fed, die of starvation in the midst of plenty. It has no capacity to originate motion.[1]

---

[1] See Flourens, Longet, and Vulpian On the Results of the Removal of the Cerebral Hemispheres in Pigeons. See, also, Ferrier, Functions of the Brain, chap. iv., and Carpenter, Human Physiology, edition of 1875, pp. 696, 697.

Gentlemen, it shall not be my fault if you go from this hall without having impressed on you the distinction between the influential and automatic nervous mechanism. Next after the contrasts between the living and the not-living, and between matter and mind, that distinction is the most important and the widest in biology. These three colossal distinctions all not only inhere, but co-inhere, in the very substance of the science of the relations of matter and mind. These are the sublime peaks of biology; and on them, in clear days, whoever would know the landscape of modern philosophy and of religious science must wander with the best telescopes well used, and pace to and fro, and be alone, and sometimes kneel.

Perfectly coincident with metaphysics is physiology, whenever the two speak on the same point. Physiology shows us two kinds of nervous activities,—one automatic, one influential—I might say volitional and responsive, but I anxiously avoid merely technical terms when the use of them is not necessary. I adopt the phraseology of Draper, "influential and automatic," rather than the phraseology of Carpenter, "volitional and responsive," because "influential" is a wider word than "volitional." I suppose that the will does originate muscular action.[1] But the will is not the whole soul. I believe that every part of the soul is "influential" on what is called the influential nervous arc. Every finger of the invisible musician who wears Gyges' ring, and which we call the soul, touches some point of this board of whitish gray keys. I will not name the activity of the whole set of fingers on this board by that of the thumb merely. To call this whole list of activities volitional would be to name but the thumb, when we have reason, imagination, emotion, all acting more mysteriously by far than the swiftest motion of your Ole Bull's Norwegian fingers on the strings of his magical instrument. Keep, then, this distinction between the influential and the automatic before your mind; remember that volitional and responsive are other words for the same things, and you will find that the great contrast between matter and mind, which is so prominent in metaphysics, is equally prominent in physiology.

I hold, that in the divine language in matter, as well as in mind, there is not an empty word, syllable, letter, space or point. By and by the time will come when everything in the universe of forms, as well as in that of forces, will be found to be significant,—doubly, trebly, quadruply, infinitely. It is safe to maintain, that this great distinction in the body between the automatic and the influential, is a thing meant to indicate to us the contrast between necessity and freedom, fate and choice. So are we woven by the bioplasts, that a part of our actions are responsive to physical, and a part responsive to spiritual stimulus. Dr. Carpenter affirms in so many words, that, in the nervous mechanism, "the *vesicular* substance has for its office to *originate* changes which it is the business of the *fibrous* to *conduct*."[2]

---

[1] Carpenter, Mental Physiology, American edition, pp. 378, 386, 391, 418.
[2] Human Physiology, edition of 1875, p. 587: see, also, pp. 694, 713, 752.

"The will," he teaches, "is constantly *initiating* movement. The distinction between *voluntary* and *involuntary* movement is recognised by every physiologist."[1]

It is Carpenter's theory that consciousness is located in the sensory ganglia, which lie immediately between the influential and the automatic arcs, and that just as an outward physical impulse may be transmitted upward through the automatic nerves to this sensory centre, so an impulse originated by pure spirit in the cerebral hemispheres may be transmitted downward to the seat of consciousness. We know what the nerves of the external senses are ; but Reil and Carpenter very significantly call the highest influential mechanism *the nerves of the internal senses*. As the automatic nerve touches light, so the influential soul. Mysterious beyond comment is this physical contrast when regarded as a first letter in the alphabet of philosophy. That part of a tree which is below the soil is not more different from that which is above than the automatic is different from the influential nervous mechanism. A ship below the water-line is adapted to the water, and above that line to the air ; but the sails and rudder are not more palpably adapted to different agents than the automatic and the influential nervous arcs in man. As well as we know that a sail is inert without wind, we do know that this upper nervous arc is inert without soul. As from the structure of the sail we might infer the nature of wind, so, from that of the inert mechanism of the brain, Draper and Lotze and Beale and Carpenter say we may infer that of the viewless spiritual force which beats on it.

What can prove to us that the upper arc of the nervous system has that behind it which has power to originate motion, unless it be the fact that the removal of that arc takes away all power in the animal to originate motion ? There is the effect ; and it ceases when the cause ceases. I ask only that you should apply here the stern law of Newton, that, where cause and effect are conjoined, the taking away of the former produces the cessation of the latter. We take away the cerebral hemisphere of the fish, the frog, the pigeon, the rabbit ; and the animals invariably become mummies from the loss of all power of originating muscular movements.

To summarise, then, a crowded discussion, let me in the name, not of Draper simply, but of Beale, of Carpenter, of Ferrier, of Lotze, of Frey, of Stricker, of Kölliker, of Wundt, and of Helmholtz, affirm,—

1. In the absence of the cerebral hemispheres, the lower nervous centres, of themselves, are incapable of originating active manifestations of any kind.

2. An animal in possession of the cerebral hemispheres exhibits a varied spontaneity of action.

3. Very palpably this is not conditioned by present impressions on the organs of sense.

4. The lower nervous centres, if they are taken alone, are concerned in automatic or responsive actions only.

---

[1] Mental Physiology, pp. 414, 379 ; see, also, On the Control of Habit by the Will, pp. 366, 367 ; On its Directing Power, pp. 386-391 ; and On its Determining Power, pp. 423-428.

5. The power of self-conditioned activity the hemispheres alone possess.

All great physiological facts reach as far into philosophy as they do into physiology. May I state, under appeal for correction, that theology in our times has a physiological side? I am perfectly amazed at the feeling that many have, that a specialist in religious science has no right to look into physiology. Why, every student of religious science must be more or less a specialist in philosophy: and philosophy is now built, not only on the investigation of consciousness, but on physiology. At Andover yonder, in the course, say, of a crowded year given to religious truth as a system, fully three months are devoted to what is called natural theology; and all the six lectures, and often more a week, turn on philosophy largely, and I had almost said exclusively. Till the existence of God and of the soul is demonstrated, religious science does not take up the topic of biblical evidence. She does take it up at last, but with an arm of resistless strength, when at last she comes to the close of natural theology, and enters on revealed. Andover, like New Haven, like Princeton, like Edinburgh, like Oxford and Cambridge, like Heidelberg, Halle, Leipzig, and Berlin, begins with axioms, with self-evident, first truths, asking no man to believe more than what Aristotle laid down as incontrovertible, self-evident, necessary, axiomatic. On the basis of that adamant, having proved the existence of God and of the soul, religious Science finds herself in an attitude to ask, What are the relations between the two? There is a God, and there is a soul; and it must be, in a universe made on a plan, that there are relations between the two; and that these relations do not depend on count of heads, or clack of tongues. The universe must have conditions of salvation in it if it is made on a plan. Religious science springs out of the universality of law. If there is a soul, and the soul is made on a plan, if there is a God who is all order and all holiness, then it is incontrovertible that there are natural conditions of salvation. What is salvation? Let us have a definition. Salvation is *permanent deliverance from both the love of sin and the guilt of sin.* It must be, that, in a universe in which we can demonstrate the existence of a living God and a living soul, conditions of freedom from the love of sin and from the guilt of it exist, that you and I cannot change by ignoring them, or voting them up or down. The government of the universe is not elective. Therefore, it is fitting for us to begin with demonstrating axiomatically the existence of God and the existence of the soul, in order that we may go forward and learn from the plan of the two what must be the natural conditions of their harmony.

Religion a science? Yes, assuredly; for science is simply *a body of established truth, or systematised knowledge, reached by the application of the scientific method, that is, by definition and induction.* By these processes, which religious science invented, she undertakes to investigate the activity of the highest zones of man's being, to establish right conduct upon the nature of things, to ascertain the contents of both natural and revealed truths, to illustrate, in short, by all that can be known to man, the relations between the soul and its Author. A science ? Yes, certainly ; a result of the use of the scientific method ; and not only as much a science as any other, but a science as much more than any other as a view from the top of the dome of St. Peter's is a greater outlook than the view from any slit called a window.

You say that only a brick-maker can understand architecture. Well, I cannot make brick ; but it has been my speciality for the last ten years to study logical, physiological, metaphysical, theological, and ethical architecture. It is trite beyond measure to say, although some sceptics seem never to have heard, that it is the duty of every theological student to know with uncommon thoroughness logic and metaphysics, and the chief results of the most advanced physiological as well as of the latest exegetical research. I should consider myself unfit to hold up my rushlight before religious truth anywhere, if I had not given myself to these topics for years, not only under the best guidance, but with the freest spirit.

Michael Angelo never learned to make a brick ; he was not skilful as a plumber : but he had some knowledge of architecture. I am willing to compare with Michael Angelo's knowledge of material architecture that knowledge of logical and philosophical architecture which belongs in our age to some teachers of religious science in Germany. A man may be an architect, although he is not a carpenter, and cannot fell a tree skilfully, or hew a stone, or unroll lead on a roof. There may be in a man sound judgment as to architecture, although he knows nothing about making brick. I revere specialists, and am not underrating them ; but very plainly the relation of all minor specialists to philosophy is that of the contributors of material to the architects of your St. Peter's or your Milan Cathedral. From all sides material comes to the architect. Each specialist guarantees his own material ; but the architect, by all the tests known to man, is to find out what are good and what are bad brick, timbers, granite, and marble ; and, whenever the sciences agree what materials are good, it is our business to build with them the temple of religious thought. We have a right to do this if we understand architecture.

A specialist is undoubtedly a king of research in his own field ; but what if that field embraces only molluscs, or scarabea, or the dative case ? A specialist may have a wide field. Who is a specialist ? I affirm that your Michael Angelo is a specialist as well as your mere brickmaker and plumber. When the minor specialists assume an arrogant attitude toward the greater, I am always reminded of the stone-cutters I conversed with in Story's studio at Rome. "We made this Cleopatra," said they ; " we produced this Sybil ; " and so through twenty resplendent works of art. And then the stone-cutters added, as a matter of small moment, "Our modeller, Mr, Story, is upstairs." Even Ernst Häckel insists upon it [1] that the narrowness of outlook of specialists in physical science, and their inadequate philosophical training, is the worst mischief of our

---

[1] History of Creation, vol. ii. p. 349.

modern scientific discussion. Do not think that I speak from prejudice in the assertion that there is no profession, unless it be the legal, better trained in logic and philosophy than the ministerial. I am aware that I am speaking before an audience containing many scholars, and I am anxious never to violate courtesy here toward learning of any kind; but I do not know where, in a course of medical instruction, any physician gets that merciless drill in logic which is necessary in any adequate theological or legal professional preparation and career. I do not know where any man studying merely with the microscope and scalpel and retort obtains that kind of literary and logical and philosophical training which belongs of necessity to the law and theology. This has been so in all ages, though we undoubtedly have made mistakes. No doubt we have sometimes taken brick that were poorly baked; and I think that is our chief trouble to-day.

In justification of the five propositions thus far discussed, let me ask you to listen to Professor Ferrier, indorsed now by Carpenter and Dalton in standard text-books of science. "One fundamental fact seems to be conclusively demonstrated by these experiments, viz., that in the absence of the cerebral hemisphere, *the lower centres of themselves are incapable of originating active manifestations of any kind*. An animal with brain intact exhibits a varied spontaneity of actions, *not, at least, immediately conditioned by present impressions on its organ of sense*. When the hemispheres are removed, all the actions of the animal become the immediate and necessary response to the form and intensity of the stimulus communicated to its afferent nerves. Without such excitation from without, the animal remains motionless and inert. It is true that some of the phenomena which have been described would seem to be opposed to this view; but they are so in appearance only, and not in reality. . . . Hence *the phenomena manifested by the different classes of animals, after ablation of the hemispheres, admit of generalisation under the law that the lower ganglia are centres of immediate responsive action only, as contradistinguished from the mediate or self-conditioned activity which the hemispheres alone possess.*" [1]

Although, from the course of his education, Ferrier might be expected to lean toward Bain's philosophy, he cannot be accused of crudeness while he maintains that the distinction between matter and mind is as clear in physiology as in metaphysics. He does that in this very significant statement of facts from a physiologist's point of view; and this to-day is the freshest word on our theme : "That the brain is the organ of the mind, and that mental operations are possible only in and through the brain, is now so thoroughly well established and recognised, that we may, without further question, start from this as an ultimate fact. But how it is that molecular changes in the brain-cells coincide with modifications of consciousness, how, for instance, the vibrations of light falling on the retina excite the modification of consciousness termed a visual sensation, is

---

[1] Ferrier, Functions of the Brain, pp. 40, 41.

a problem which cannot be solved. We may succeed in determining the exact nature of the molecular changes which occur in the brain-cells when a sensation is experienced; but this will not bring us one whit nearer the explanation of the ultimate nature of that which constitutes the sensation. The one is objective and the other subjective: and neither can be expressed in terms of the other. *We cannot say that they are identical, or even that the one passes into the other, but only, as Laycock expresses it, that the two are correlated.*" [1]

Just here I must fulfil my promise to refer to a courteous question asked me in print [2] by a gentleman who thinks that "chemical force and vital force are cognate." That is his language; and by it I understand him to mean that the one is kindred in origin with the other. Certainly he does not hold himself in such an attitude in this article that he can be exonerated from the grave charge that he disagrees with Ferrier, when the latter teaches, as Tyndall affirms also, that these molecular activities "*cannot be made to pass into*" mental activities. Speaking of the effect of "tea and coffee and phosphorated food in oiling the wheels of the mind," this Boston writer says, "Such agents develop chemical force without question : this force, to the best of our knowledge, accelerates the wheels of life, and it is every way proper to suppose that, doing thus, it is analogous to the force which sets the wheels going ; or, in short, that chemical force and vital force are cognate." He then goes on to affirm that the "impressions" coming from different quarters "are to the individual the representative of the universe, and that it may be said that in this way the universe is each man's tutor, and *forms* his soul." Gentlemen, that is materialism.

Let us test this typical statement by a parallel case. The reasoning may be summarised in three propositions : (1) Chemical force accelerates the wheels of life; (2) Therefore it is analogous to the force which sets the wheels of life in motion ; (3) Therefore chemical and vital forces are cognate. Now let us parallel that reasoning, point for point, for the sake of clearness. The strong current in the Merrimack or Charles River accelerates the motion of the rower in his boat. It is, therefore, every way proper to suppose that the force of the current is analogous to the force which sets the oars in motion.

I beg you to be courteous, gentlemen. This Lectureship has but one motto, "The clear, the true, the new, the strategic." I do not first seek orthodoxy ; I seek first clearness. A man who sets before himself even truth as the first object is likely to make truth only the synonyme for his own opinion. Let us seek first clearness, whether the heavens stand or fall.

To proceed, then : the force in the current accelerates the motion of the rower in his boat : therefore it is every way proper to suppose that it is analogous to the force that sets the oars in motion ; and therefore the force of the current and the force that moves the oars are cognate.

But this is not all ; for, to make the parallel complete, we must

---

[1] Ferrier, Functions of the Brain, pp. 255, 256.   [2] Daily Advertiser, Nov. 29, 1876.

assert that the force that moves the currents and the force that moves the oars are cognate in such a sense, that, when all things are fairly stated, it must be conceded that the force that moves the currents " forms " the force which moves the oars.

Undoubtedly the rower on the river is aided by the currents, and so, undoubtedly, is the rower called life aided by currents of purely physical force moving through the living organism; but to say that from this fact we must conclude that the two forces are cognate, is no more unreasonable in the former case than in the latter.

This gentleman thinks, that, at one point, I make a leap in my proof; but I never leaped across the difference between the current in the river and the force that moves the oars.

I need not mention in detail the reasoning in an earlier paragraph of this criticism; for the concessions made to me there destroy the criticism, and the whole falls when the word "cognate" falls. The gentleman says it is "force" which moves that portion of the brain which will not react under electrical stimulus. I say it is *"force,"* but not physical force; for this, as Ferrier says, cannot be shown to pass into mental force. This gentleman's reasoning to prove that it does so pass proves astoundingly too much. The force, too, must be one adequate to account for the effect produced.

When he grave assertion is made that the bellows yonder accelerates the action of the organ, and that, therefore, it is perfectly proper to suppose that its force of rough wind is of the same character with the will of the musician whose fingers touch the keys, and that, therefore, *the musician was blown out of the bellows*, we come to a vivid view of the logic of materialism.

You put me into a bad mood, gentlemen. I have heard that hypotheses are allowable up to a certain point, but that there does come a time in logic when there must be an end of hypotheses. De Morgan, in his logic, tells a story of a servant who was to prepare a stork for dinner for his master. But the servant had a sweetheart; and, to gratify her, he cut off a leg of the stork after it had been cooked, and put the mutilated bird upon the table of the nobleman. When dinner was served, the nobleman called the servant to the door of the feasting-hall, and said, "How does it happen that this stork has but one leg?"—"Why, sir," was the hypothesis used in answer, "a stork never has but one leg." No more was said in the presence of the company; but the next day, before the nobleman dismissed his servant, he thought he would see what further hypothesis the man would offer. So he took his servant into the grounds of the castle, and showed him the storks standing there. "See," the nobleman said, "each stork has two legs."—"But look again," said the servant, "each stork has really now but one;" and surely each was standing, after the manner of this bird, on one. But the nobleman shouted to the birds with a frightening gesture, "Off, away!" and each stork ran away with two legs. "Yes," said the servant, who did not lack hypothesis; "but, yesterday you did not say, 'Off and away' to that stork on the table." There must at some point be an end to hypothesis; but materialism saves itself by saying, "Off and away!"

to the baked stork. Why, the poor grave-digger in Hamlet knew better than that; for he was no materialist. "Who is to be buried here?" said Hamlet; and the fool answered—

> "One that was a woman;
> But, rest her *soul*, she is *dead*."

At our present point of view, we need only name the propositions which flow from the latest physiological research :—

6. Molecular motions in the nervous system are now definitely known to form in all cases a closed circuit.

7. They cannot, therefore, be said to be identical with mental activities.

8. They are only parallel with them.

9. They are demonstrably not transmuted into mental activities, but only correlated with them. Parallelism is not identity : the keys in motion are not the music of your organ.

10. Materialism, therefore, fails under the microscope of physiology, and it fails equally under a strict application of the law of causation.

The externality of the soul to the nervous mechanism is just as well known in relation to the upper keyboard or influential arcs, as the externality of your fingers to the lower keyboard or the automatic arcs, is known in these experiments with the frog and the pigeon, the fish and the rabbit. You know how those motions in the lower keyboard are produced. You know, therefore, how those in the upper are started. Matter did not start them there. Matter does not start them here. Mind starts them here. Mind starts them there. We are conscious in ourselves of power of choice, and that inner witness must be combined with the testimony that comes from the scalpel and the microscope, to show that the power of self-direction does not originate in matter.

How the unextended substance, mind, can act upon the extended substance, matter, is a mystery ; but to affirm that it does so involves no self-contradiction. What is a mystery ? Something of which we know *that* it is, though we do not know *how* it is. What is a self-contradiction ? An inconsistency of a proposition with its own implications. That mind moves matter, we know. How it does it, we know not. Sir William Hamilton,[1] in his efforts to solve this mystery was anxious that even what is called mesmeric force should be investigated ; and he and many other acute minds have asked whether it may not be within the power of the human will to influence another human will across the street, across the city, across a continent. In the name of exact science, many seek to-day to know whether by possibility human will may not, in some cases, make matter move by willing to do it. I hold no strange theory on this theme ; I am shy to my fingers' tips of even the conclusions of Carpenter concerning it. But will you not allow me, in the name of Sir William Hamilton's curiosity, and in that of President Wayland of Brown University, to use, merely as illustration, this presumed power of the human will to

---

[1] Professor Veitch, Memoir of Sir William Hamilton, p. 154.

move matter without contact through other matter? If you conceive that as possible, and fairly within natural law, then natural law itself becomes the magnetisation of all matter by the influence of one Omnipresent Will, in which is no variableness nor shadow of turning. As our wills play upon the keyboard of the influential human nerves, so Omniscience and Omnipresence, magnetising all worlds and their inhabitants, play upon all infinities and eternities. The connection of the Divine Will with matter may be thus obscurely revealed to us by that of the human will with matter. Each is a mystery; but, if these two are kindred mysteries, the universe is one, and man's passion for unity in science is satisfied. Matter is an effluence of the Divine Nature, and so is all finite mind, and thus the universe is one in its present ground of existence and in the First Cause. In a better age, Science, lighting her lamp at that Higher Unity, will teach that, although He, whom we dare not name, transcends all natural laws, they are, through His Immanence, literally God, who was, and is, and is to come. Science does this already for all who think clearly.

XII.

EMERSON'S VIEWS ON IMMORTALITY.[1]

" Ψευδηγορεῖν γὰρ οὐκ ἐπίσταται στόμα
τὸ δῖον, ἀλλὰ πᾶν ἔπος τελεῖ."
ÆSCHYLUS, *Prometheus Bound*, 1031.

" Οὐκ ἂν τάδ' ἔστη τῇδε, μὴ θεῶν μέτα."
SOPHOCLES, *Ajax*, 950.

PRELUDE ON CURRENT EVENTS.

WHICH city has the greater right to an attitude of intellectual haughtiness, Boston or Edinburgh? In preparation for all inspired work in poetry and art, and, much more, in religion, it is necessary to make the palms of the hands clean and to shake off them the glittering, stout vipers,—intellectual pride, vanity, and self-sufficiency. Has Edinburgh shown a greater decision and skill than Boston in dislodging these wreathing reptiles from her fingers, as Paul shook off the serpent on Melitus, feeling no harm? Is Edinburgh really the equal of Boston in culture? Where is there in this city a better metaphysician than Sir William Hamilton or Dugald Stewart? Who here has advanced exact science more than Black, or Playfair, or Sir David Brewster? Is there a better political economist here than Adam Smith, the author of "The Wealth of Nations"? Have we better historians than Hume and Robertson? Is there any rhetorician here likely to be more influential than Hugh Blair? Have we a painter superior to Sir John Leslie, a more delightful essayist than Thomas De Quincey, a better writer on ethics than Sir James Mackintosh? What literary name have we, on the whole, superior to that of Walter Scott? Can Boston produce the equal of John Knox or Thomas Chalmers? What periodical of the same class have we better than "Blackwood's Magazine," as edited by a Lockhart and a Wilson? What quarterly have we here in Boston more famous than "The Edinburgh Review," with Francis Jeffrey, and Sidney Smith, and Horner, and Macaulay, and Brougham behind it? This Edinburgh, true to the deepest inspirations of conscience in her Scotch heart and intellect, knelt down lately on the shore of the North Sea, and was willing to have her devotions led by an American evangelist; and shall Boston, on this Puritan and Pilgrim shore, stand stupidly stiff when asked to kneel?

Dickens wrote in his last years, that he regarded a Boston audience as next to an Edinburgh audience, but that this was a high compliment to Boston; for he regarded an Edinburgh audience as perfect.

What if Boston in 1877 should receive, as well as Edinburgh did in 1874, evan-

---

[1] The fifty-seventh lecture in the Boston Monday Lectureship, delivered in Tremont Temple.

gelists thrice more emphatically approved by experience now than they were then? What if we should put ourselves as thoroughly as Edinburgh did herself into the attitude of a telescope focused on the sun of religious truth, and ready, therefore, to cause an image of the sun to spring up in the chambers of the instrument? We are proud of our lenses: are we willing to adjust them? *Once adjusted, even poor human lenses, by fixed natural law, may draw down a star or a sun into the soul; and, although the light is from above, the adjustment is our own.* Are we willing to bring the axis of adjusted, spiritual, telescopic thought in Boston into complete coincidence with the line of the keenest rays of conscience, and of self-surrender to God, and see what the effect will be in the starting-up within us of a light otherwise unattainable, and hot enough to burn up our temptations,—hot enough to purge whatever of politics, or commerce, or social life is held in the focus of the rays,—hot enough to sear the wings of the dolorous and accursed scepticisms which flutter not through the Boston noon, but through the Boston dusk, and endeavour yet to build homes for themselves in last year's birds' nests, like Paine's forgotten books, and Parkerism, and small philosophy, and free religion and materialism?

Edinburgh, when Mr. Moody came to that city, avoided a division of her Christian forces. Half a score of churches could not hold the audiences; but there was no lack of trained minds and hearts ready to converse with the religiously irresolute face to face. To bring those who have not surrendered to God face to face with those who have, and to let the two sets of minds act and react upon each other in personal hushed conversation, religious study, and prayer, is one of the highest blessings to both, and perhaps the most effective human instrumentality known to man for the diffusion of personal religion. I have seen men and women go into such conversation shiveringly as babes into a bath, and come out with foreheads white and eyes like stars. Face-to-face conversation between the converted and the unconverted is everywhere the chief measure to be taken for the religious culture of *both*. The secret of Mr. Moody's great usefulness is in a combination of three things,—his total and immeasurably glad self-surrender to God; his fervid oratory, alive in every part with biblical truth, practical sagacity, and fathomlessly genuine consent to conscience; and his most uncommon good sense in organising religious effort in those forms which bring the converted and the unconverted face to face in conversation, biblical study, and prayer.

A power not of man springs up when the religiously resolute and the religiously irresolute converse and kneel together in the holy of holies of human experience, a Divine aroma breathed upon the two from the open Scriptures between their eager faces. These inquiry-meetings, this organisation of lay religious effort, this putting the unrepentant face to face with the converted, this kneeling together of those who are right with God and those who wish to be, is the secret, I think, of the chief religious power in the long course of the evangelist's work.

Edinburgh was willing, with all her haughtiness, to enter into that style of religious effort. Professor Blaikie says that the sacred songs which filled the meetings are at this day better known in Scotland than Burns' poems. In a call issued to all Scotland from Edinburgh, nearly all the professors of the University of Edinburgh are represented. There were in the list of signatures the names of Professors Calderwood, Balfour, Blaikie, Charteris, MacGregor, and Crawford, side by side with those of Hanna and Duff, Scott Moncrieff, and Horatius Bonar. There is hardly a circle of culture in Edinburgh that was not proud to be represented in the lists of persons engaged in the endeavour to carry the truth to all portions of society. Will Boston do anything like this? Will Harvard University do what Edinburgh University was proud to do to carry men on a vigorous current of calm thought into self-surrender to God? I wish to speak with due reverence of this city; but I am not of the opinion that Boston is entitled to more intellectual renown than Edinburgh; and yet, in Edinburgh, the students came out by thousands to hear religious truth, and to have a personal application made of it to themselves, not altogether by the evangelist, but by the spirit of the time. You remember that on one occasion the students of Edinburgh came together in the Free Assembly Hall, and so filled it, that Mr. Moody was obliged to speak to an immense gathering in the quadrangle, while Mr. Whyte, successor to Dr. Candlish, and Professor Charteris, conducted the services within. Around the platform were professors from nearly all the departments of the university, and from the Free Church and College, and nearly two thousand students. This was a more significant scene than that when Gladstone sat on the platform in London.[1]

Edinburgh is looking upon Boston; London watches this city; Glasgow, Liverpool, Philadelphia, New York, Chicago, ask what Boston will do to bring herself into an attitude in which God can walk up and down our streets as He has walked up and down the streets of other cities. Who will prepare the way for the triumphal procession, not of any sect, but of all Christian truth? In Chicago, the other day, a young man who had stolen some thousands of dollars confessed his sin to the person with whom he conversed in an inquirer's room, and of his own accord went to the penitentiary. Over and over again it has happened in these meetings that men guilty of unreportable deeds have confessed them, and have begun new lives with that emphasis of sincerity which is given by voluntarily taking witnesses to utterly unspeakable guilt. Is this excitement? It is Almighty God in conscience. Professor Dorner I heard say once in Berlin University, "The truth is, gentlemen, not so much that man has conscience as that conscience has man." Your Emerson says men cannot love Goethe, because he was incapable of surrender to the moral sentiment. Is Boston ready to give herself up to that sentiment in such a manner that she shall not only know that she has conscience, but allow conscience, and God who is behind it, to have her?

THE LECTURE.

As light fills, and yet transcends, the rainbow, so God fills, and yet transcends, all natural law. According to scientific Theism, we are equally sure of the Divine Immanency in all Nature, and of the Divine Transcendency beyond it. Pantheism, however, with immeasurably narrow horizons, asserts that natural law and God are one; and thus, at its best, it teaches but one-half the truth; namely, the Divine Immanency, and not the Divine Transcendency. Christian Theism, in the name of the scientific method, teaches both. While you are ready to admit that every pulsation of the colours seven in the rainbow is light, you yet remember well that all the pulsations taken together do not constitute the whole of light. Solar radiance billows away to all points of the compass. Your bow is bent above only one-quarter of the horizon. So scientific Theism supposes that the whole universe, or finite existence in its widest range, is filled by the infinite Omnipresent Will, as the bow is filled with light; and this in such a sense that we may say that natural law is God, who was, who is, and who is to come. In the incontrovertible scientific certainty of the Divine Immanency, we may feel ourselves transfigured as truly as any poetic pantheist ever felt himself to be when lifted to his highest possible mount of vision. But, beyond all that, Christian Theism affirms that God, knowable but unfathomable, incomprehensible but not inapprehensible, billows away beyond all that we call infinities and eternities, as light beyond the rainbow. While He is in all finite mind and matter, as light is in the colours seven, He is as different from finite mind and matter as is the noon from a narrow band of colour on the

---

[1] Dr. John Hall and G. H. Stuart, The American Evangelists, p. 51.

azure. Asserting the Divine Transcendency side by side with the Divine Immanency, religious science escapes, on the one hand, the self-contradictions and narrowness of pantheism, and attains, on the other, by the cold precision of exact research, a plane of thought as much higher than that of materialism as the seventh heaven is loftier than the platform of the insect or the worm.

It would be very Emersonian to differ from Emerson. His mission, according to his own statement, is to unsettle all things. It is common to hear the acutest readers assert that his writings have no mental unity. The poet Lowell thinks that sometimes Emerson's paragraphs are arranged by being shuffled in manuscript; and the best British criticism[1] says, "They are tossed out at random like the contents of a conjuror's heart." But is there no point of view from which the Emersonian sky,

"With cycles and with epicycles
Scribbled o'er,"

may be seen to have within it a comprehensible law? Before Hegel, Emerson's master, became obsolete or obsolescent in Germany, no doubt Emerson was a pantheist; but I cannot explain by any form of pantheism the later motions of some stars in his pure, soft azure. You may prove that he is more poet than philosopher, more seer than poet, more mystic than seer; and yet the surety in the last analysis is, that he is more Emerson than either. *Individualism held firmly, pantheism held waveringly*, are to me the explanation of the bewildering and yet gorgeous motions of the constellations in his sky. Mr. Frothingham acutely says that Mr. Emerson's place is among poetic, not among philosophic minds.[2] It is not Emersonian to wince under philosophical self-contradiction; but it is Emersonian to writhe under the remotest attempt to cast on individualism so much as the fetter of a shadow.

Loyalty to the Over-Soul is Emerson's supreme mood. Whether it lead to philosophic consistency or not, is to his scheme of thought an empty question. Whatever shooting-star streams at this instant across the inner sky of personal inspiration is to be observed, and its course mapped down, even if it move in a direction opposite to that of the last flaming track of light noted there. What if the map at last show a thousand tracks crossing each other? Are they not all divine paths? Are they not to be all included and explained in a sufficiently wise philosophy? The point of departure of all the shooting-stars in Emerson's sky is the constellation Leo. All his metaphysics he is ready to abandon at any moment, if the loftier movements of the soul as its exists in himself come into conflict with his philosophy. He utters whatever the Over-Soul seems to him to say, whether in harmony with previous deliverances or not. He is a pantheist, but not a consistent pantheist: he is an idealist, but not a consistent idealist: he is a religious mystic, but not a consistent

---

[1] Encyc. Brit., 1875, art. "On American Literature."
[2] Transcendentalism in New England, 1876, p. 236.

mystic: *he is an individualist, mapping his own highest inner self, or as he would say in pantheistic phrase, mapping God.* The Over-Soul comes to consciousness only in man. In the transfigured work of tracing on the page of literature all gleams of light in the Over-Soul in Emerson, he is consistent with himself, and in this only. A maker of maps of the paths of shooting-stars is Emerson; and he is more devout than any astronomer intoxicated with the azure. Sit in the constellation Leo, if you would understand the Emersonian sky.

A brilliant and learned volume by a revered preacher of this city [1] contains the most luminous analytical proof that a pantheistic trend sets through Emerson's writings as the Gulf-current through the Atlantic. But Emerson often proclaims his readiness to abandon pantheism itself, if the Over-Soul seems to command him to do so. In the whole range of his often self-destructive apothegms, I find no single sentence so descriptive of his position as a fixed individualist and a wavering pantheist as this:—

"In your metaphysics you have denied personality to the Deity; yet, when the devout motions of the soul come, yield to them heart and life, though they should clothe God with shape and colour. Leave your theory, as Joseph his coat in the hand of the harlot, and flee." [2]

Whoever would come to the point of view from which all Emerson's self-contradictions are reconciled must take his position upon the summit of individualism, and transfigure that height by the thought that there billows around it what we call God in conscience, and what Emerson calls the Over-Soul. In the loftiest zones of human experience there are influences from a Somewhat and Someone that is in us, but not of us; and Emerson is so far pantheistic as to hold that this highest in man is not only a manifestation of God, but God, and the only God. Therefore he is always in the mount. His supreme tenet is the primacy of mind in the universe, and I had almost said the identity of the human mind with the Divine Mind. As the waves are many, and yet one with the sea, so to pantheism finite minds and the events of the universe are many, and yet one with God. As the green billows that dash at this moment on Boston Harbour bar, and cap themselves with foam, are one with the Atlantic, so you and I, and Shakspeare, and Charlemagne, and Cæsar, and the Seven Stars, and Orion, are but so many waves in the Divine All. The ages, like the soft-hissing spray, may take this shape or that; but they all come from one sea. Every wave is an inlet to the sea, and to all of the sea. "There is," says Emerson, "one Mind common to all individual men. Every man is an inlet to the same, and to all of the same." [3] "*The simplest person, who in his integrity worships God, becomes God.*" Eight generations of clerical descent are behind Emerson's unwavering reverence for the still small voice: one generation of now almost outgrown German thinkers is behind his wavering

---

[1] Rev. Dr. Manning, Half Truths and the Truth, 1872.
[2] Emerson, Essays, vol. i. p. 50.   [3] Essay on History.

reverence for pantheism. Would he only assert, side by side with the Divine Immanence, the Divine Transcendency, we might call him a Christian mystic, where now we can only call him a teacher of transfigured pantheistic individualism.

Pantheism denies the personal immortality of the soul. To pantheism, death is the sinking of a wave back into the sea. We shall find, however, that Emerson, true to his central tenet of hallowed individualism, has again and again asserted the personal immortality of the soul, and never denied it in reality, though he has often done so in appearance.

When, in 1832, Mr. Emerson bade adieu to his parish in this city, he used, as on every occasion he is accustomed to use, memorable words. "I commend you," the last sentences of his letter to that parish read, "to the Divine Providence. May He multiply to your families and to your persons every genuine blessing; and whatever discipline may be appointed to you in this world, may the blessed hope of the resurrection, which He has planted in the constitution of the human soul, and confirmed and manifested by Jesus Christ, be made good to you beyond the grave! In this faith and hope I bid you farewell."[1] These are wholly unambiguous words.

You say that Emerson never has asserted, since 1832, the personal immortality of the soul; but what do you make of certain almost sacredly private statements of his to Fredrika Bremer? That authoress, whose works Germany gathers up in thirty-four volumes, came out of the snows of Northern Europe, and one day found Mr. Emerson walking down the avenue of pines in front of his house, through the falling snow, to greet her. Day after day they conversed on the highest themes. Months passed while Fredrika Bremer was the guest of Boston; and toward the end of the lofty interchanges of thought between these two elect souls, there occurred what Fredrika Bremer calls a most serious season. One afternoon in Boston, with all the depth of her passionate and poetic temperament, she endeavoured to convince Emerson that God is not only in all natural law, but that He transcends it all; that He demands of us perfection; and that, therefore, as Kant used to say, we must expect personal immortality or opportunity to fulfil the demand; that religion is the marriage of the soul with God; and that the idea that God is objective to us, and that our souls may come into harmony with His, a Person meeting a person, is vastly superior, as an inspiration, to any pantheistic theory that all there is of God is what is revealed to us in the insignificant scope of our faculties. She endeavoured, in the name of lofty thought, to show the narrowness of pantheism at its best. The interview was serious in the last degree; and Fredrika Bremer says that Emerson closed it with these words, "I do not wish that people should pretend to know or believe more than they really do know and believe. The resurrection, the continuance of our being, is granted: we carry the pledge of this in

---

[1] R. W. Emerson, Letter dated Boston, Dec. 22, 1832, quoted in Frothingham's Transcendentalism in New England, 1876, p. 235.

our own breast. I maintain merely that we cannot say in what form or in what manner our existence will be continued."[1]

Transcendentalism in New England was marked by a bold assertion of the personal continuance of the soul after death. "The Dial" always assumed the fact of immortality. "The transcendentalist was an enthusiast on this article," Mr. Frothingham says; and Mr. Emerson's writings, he adds, were "redolent of the faith." Theodore Parker thought personal immortality is known to us by intuition, or as a self-evident truth, as surely as we know that a whole is greater than a part. It must be admitted that New-England transcendentalism caused in many parts of our nation a revival of interest and of faith in personal immortality.[2] Mr. Emerson was the leader of New-England transcendentalism.

But you say, that since 1850, Emerson has changed his opinion; and yet, if you open the last essay he has given to the world, that on "Immortality," you will read, "Everything is prospective, and man is to live hereafter. That the world is for his education is the only sane solution of the enigma. . . . The implanting of a desire indicates that the gratification of that desire is in the constitution of the creature that feels it. . . . The Creator keeps His word with us. . . . All I have seen teaches me to trust the Creator for all I have not seen. Will you, with vast cost and pains, educate your children to produce a masterpiece, and then shoot them down?" What do these phrases amount to, taken in connection with the two earlier passages which I have cited, and which assuredly assert personal immortality? "All sound minds rest on a certain preliminary conviction, namely, that, if it be best that *conscious personal life* shall continue, it will continue; if not best, then it will not; and we, if we saw the whole, should, of course, see that it was better so. . . . I admit that you shall find a good deal of scepticism in the street and hotels and places of coarse amusement; but that is only to say that the practical faculties are faster developed than the spiritual. Where there is depravity, there is a slaughter-house style of thinking. One argument of future life is the recoil of the mind in such company,— our pain at every sceptical statement."

The "conscious personal" continuance of the soul, Emerson no more than Goethe denies. In this very essay, however, we must expect to find apparent self-contradiction; and accordingly we can read here these sentences, written from the point of view of a wavering pantheism, "Jesus never preaches the personal immortality. . . . I confess that everything connected with our personality fails. The moral and intellectual reality to which we aspire is immortal, and we only through that."

Allow me, on this occasion, to contrast arguments with *ipse dixits*, and to use only the considerations which are implied in Emerson's teachings on immortality. You will be your own judges whether the conclusion that there is a personal existence after death must follow from his

---

[1] Emerson, Conversation with Fredrika Bremer, Homes of the New World, vol. i. p. 223.   [2] See Frothingham, Transcendentalism, pp. 195-198.

premises. I shall, of course, unbraid the reasoning, and show its strands; but its braided form is Emerson's axiom, "The Creator keeps His word with us." The argument is old; and for that reason, probably, Emerson values it. It has borne the tooth of time, and the buffetings of acutest controversy age after age. In our century it stands firmer than ever, because we know now through the microscope, better than before, that there is that behind living tissues which blind mechanical laws cannot explain.

1. An organic or constitutional instinct is an impulse or propensity existing prior to experience, and independent of instruction.

This definition is a very fundamental one, and is substantially Paley's.[1]

2. The expectation of existence after death is an organic or constitutional instinct.

3. The existence of this instinct in man is as demonstrable as the existence of the constitutional instincts of admiration for the beautiful, or of curiosity as to the relations of cause and effect.

What automatic action is, you know; and an instinct is based upon the automatic action of the nervous mechanism. Who doubts that certain postures in anger, certain attitudes in fear, certain others in reverence, certain others in surprise, are instinctive? These postures are taken up by us, without reflection on our part: they are organic in origin. It is instinct for us to rest when we are fatigued, and to take the recumbent position; and we do not reason about this. The babe does it. Instinctive actions appear early in the progress of life, and are substantially the same in all men and in all times. An educated impulse does not appear early, and is not the same among all men in all times. Of course, it would avail nothing if I were to prove that the belief in immortality has come to us from education. If that belief result from an organic instinct, however, if it be constitutional, then it means much, and more than much.

4. The dulness of these instincts in a few low races, or in poorly-developed individuals, does not disprove the proposition, that admiration for the beautiful, and curiosity as to the relations of cause and effect, are constitutional in man.

5. So the occasional feebleness of the expectation of existence after death does not show that it is not an organic or constitutional instinct.

6. This instinct appears in the natural operations of conscience, which anticipates personal punishment or reward in an existence beyond death.

You desire incisive proof that we have a constitutional anticipation of something beyond the veil; but can you look into Shakspeare's mirror of the inner man, and not see case after case of the action of that constitutional expectation? Shakspeare's delineations are philosophically as unpartisan and as exact as those of a mirror. Is it not the immemorial proverb of all great poetry, as well as of all profound philosophy, that there is something that makes cowards of us all as

---

[1] Natural Theology, chap. 18.

we draw near to death, and that this something is not physical pain but a Somewhat behind the veil? Death would have little terror if its pains were physical and intellectual only. There is an instinctive action of the moral sense by which we anticipate that there are events to come after death, and that these will concern us most closely.

Bishop Butler, in his famous "Sermons on Conscience," has no more incisive passage than that in which he declares that "conscience, unless forcibly stopped, magisterially exerts itself, and always goes on to anticipate a higher and more effectual sentence which shall hereafter second and confirm its own." This prophetic action of conscience I call the chief proof that man has an instinctive expectation of existence after death. We are so made, that we touch somewhat behind the veil. As an insect throws out its antennæ, and by their sensitive fibres touches what is near it, so the human soul throws out the vast arms of conscience to touch eternity, and Somewhat, not ourselves, in the spaces beyond this life. All there is in literature, all there is in heathen sacrifice, continued age after age, to propitiate the powers beyond death, all there is in the persistency of human endeavour, grotesque and cruel at times, to secure the peace of the soul behind the veil, are proclamations of this prophetic action of conscience; yet conscience itself is only one thread in the web of the pervasive organic instinct which anticipates existence after death.

7. This instinct appears in a sense of obligation to meet the requirements of an infinitely perfect moral law.

We know that the moral law is perfect, and therefore that the moral Lawgiver is perfect.

But the moral law demands our perfection. "Therefore," said Immanuel Kant, "the moral law contains in it a postulate of immortality." Its requirement is a part of our constitution, and cannot be met in this stage of existence. It is not met here, and therefore the moral law requires us to believe in an existence after death. That is Kant's very celebrated proof; but I am pointing to it only as one thread in this organic web which we call instinctive anticipation of existence after death. Put your Shakspeare on the fear of what is behind the veil, side by side with your Kant on this anticipation of the time when we can approximate to perfection, and you will find these broad-shouldered men, in the name of both poetry and philosophy, affirming, as the postulate of organic instinct in man, that existence after death is a reality.

8. It appears in the universality of the belief in existence after death. All widely-extended beliefs result much more from organic instinct than from tradition.

9. It appears in the human delight in permanence.

10. It appears in the unoccupied capacities of man in his present state of being.

11. It appears in the convictions natural to the highest moods of the soul.

"There shines through all our earthly dresse
Bright shootes of everlastingnesse."

12. It appears in the longing for personal immortality characteristic of all high states of complete culture.

13. It appears conspicuously in Paganism itself, in the persistence of all the ages of the world in the efforts to propitiate Supreme Powers, and to secure the peace of the soul beyond the grave.

How is the force of any impulse to be measured, unless by the work it will do? What work has not this desire of man, to be sure that all will be well with him beyond the veil, not done? What force has maintained the bloody sacrifices of the heathen world through all the dolorous ages of the career of Paganism on the planet? What force has given intensity to the inquiries of philosophy as to immortality? What has been the inspiration of the loftiest literature in every nation and in all time, whenever it has spoken of avenging deities that will see that all is made right at last? How are we to explain the persistency of every age in the attempt to propitiate the powers beyond the veil, and to secure the peace of the soul after death, if not by this impulse arising organically, and existing as a part of the human constitution?

14. Nature makes no half-hinges. God does not create a desire to mock it. The universe is not unskilfully made. There are no dissonances in the divine works. Our constitutional instincts raise no false expectations. Conscience tells no Munchausen tales. The structure of the human constitution is not an organised lie. "The Creator keeps His word with us."

15. But, if there is no existence after death, conscience does tell Munchausen tales; man is bunglingly made; his constitution raises false expectations; his structure is an organised lie.

Our age has many in it who wander as lost babes in the woods, not asking whether there is any way out of uncertainties on the highest of all themes, and in suppressed sadness beyond that of tears. Small philosophers are great characters in democratic centuries, when every man thinks for himself; but lost babes are greater. There is a feeling that we can know nothing of what we most desire to know. I hold, first of all, to the truth that man may know, not everything, but enough for practical purposes. If I have a Father in heaven, if I am created by an intelligent and benevolent Being, then it is worth while to ask the way out of these woods. I will not be a questionless lost babe; for I believe there is a way, and that, although we may not know the map of all the forest, we can find the path home.

There are four stages of culture; and they are all represented in Boston to-day, and in every highly civilized quarter of the globe. There is the first stage, in which we usually think we know everything. Then comes the second stage, in which, as our knowledge grows, we are confronted with so many questions which we can ask and cannot answer, that we say in our sophomorical, despairing mood, that we can know nothing. A little above that we say we can know something, but only what is just before our senses. Then, lastly, we come to the stage in which we say, not that we can know everything, not that we can know much, indeed, but in which we are sure we can know enough for practical purposes.

Everything, nothing, something, enough! There are the infantine, adolescent, juvenile, and mature stages of culture.

16. But, so far as human observation extends, we know inductively that there is no exception to the law that every constitutional instinct has its correlate to match it.

17. Wherever we find a wing, we find air to match it; a fin, water to match it; an eye, light to match it; an ear, sound to match it; perception of the beautiful, beauty to match it; reasoning power, cause and effect to match it; and so through all the myriads of known cases.

18. From our possession of a constitutional or organic instinct by which we expect existence after death, we must therefore infer the fact of such existence, as the migrating bird might infer the existence of a South from its instinct of migration.

19. This inference proceeds strictly upon the scientific principle of the universality of law.

20. It everywhere implies, not the absorption of the soul into the mass of general being, but its personal continuance.

Your poet, William Cullen Bryant, once sat in the sweet countryside, and heard the bugle of the wild migrating swan as the bird passed over him southward in the twilight. Looking up into the assenting azure, this seer uttered reposefully the deepest words of his philosophy :—

> "Whither, midst falling dew,
> While glow the heavens with the last steps of day,
> Far through their rosy depths dost thou pursue
>   Thy solitary way?
>
> There is a Power whose care
> Teaches thy way along that pathless coast,
> The desert and illimitable air,
>   Lone wandering, but not lost.
>
> He who, from zone to zone,
> Guides through the boundless sky thy certain flight,
> In the long way that I must tread alone
>   Will lead my steps aright."
>
>   BRYANT, *To a Waterfowl.*

XIII.
ULRICI ON THE SPIRITUAL BODY.[1]

"Der Leib der Menschen ist eine zerbrechliche, immer erneuete Hülle, die endlich sich nicht mehr erneuen kann."—HERDER, *Philosophy of History.*

> "The poet in a golden clime was born,
> With golden stars above;
> Dowered with the hate of hate, the scorn of scorn,
> The love of love.
> He saw through life and death, through good and ill;
> He saw through his own soul;
> The marvel of the everlasting will,
> An open scroll,
> Before him lay."   —TENNYSON.

PRELUDE ON CURRENT EVENTS.

THIS morning, the bells of Christian churches on the continents, and of Christian vessels on the great deep, are audible to each other around the whole planet. I am not speaking rhetorically, but geographically, when I say that the Christian Church at this moment encircles the world in her arms. We forget too often what a great continent Australia is, and what a pervasive force her English language and laws may become in the lonely southern hemisphere. But Japan has forced herself upon the notice of the world of late, as the undeveloped England of the Pacific. Her great Mikado congratulated our President, only the other day, on the success of our Centennial Exhibition; and there lay behind the cordial words from the far shore just the sentiment which a Japanese high official expressed lately at Hartford, that the Christianisation of Japan is an event to be expected in the near future. The revolution in that crowded island of sensitive, ingenious men, is in the hands of the cultivated upper classes. It does not depend on count of heads or clack of tongues, and is not likely to go backward.

You say Russia and England may come into armed collision in the shadow of the Himalayas, and that the bear and the lion may fill the Cashmere vale with blood. May God avert this! But, even if they do so, it will yet remain sure, in any event, that the days of Buddhism are numbered; and that, so far as Paganism governs Central Asia, it is every year squeezed more and more nearly to its exit from life between the state necessities of Russia and England. Coming farther West, it is significant that the Suez Canal, the key to the great gate of the way to India, belongs now chiefly to Great Britain; and that, even with the Egyptian road to the East in her possession, she cannot afford as yet to take off from Constantinople an eye behind which, for eight hundred years, has rested no inconsiderable portion of authority on this planet, and which now rules a fifth part of the population of the globe.

---

[1] The fifty-eighth lecture in the Boston Monday Lectureship, delivered in Tremont Temple.

Only this morning, from under the sea, we have whispered to us by electric lips great promises by the "sick man" of the Bosphorus. The liberty of Ottomans is to be inviolable. The religious privileges of all communities, and the free exercise of public worship by all creeds, are guaranteed. Liberty of the press is granted. Primary education is compulsory. All citizens are eligible to public offices, irrespective of religion. Confiscation, statute labour, torture, and inquisition are prohibited. Ministerial responsibility is established. A chamber of deputies and a senate are instituted. These two houses, in connection with the ministry, have the initiative in framing laws. General and municipal councils are to be formed by election. The prerogatives of the Sultan are to be only those of the constitutional sovereigns of the West.

In 1453 Islam crossed the Bosphorus with a bound, for the leprosies of its social life had not yet had time to unstring its nerves. Its own poisons have made it now little more than unspeakably flaccid flesh, without a soul. Its promises are very empty. But this time, as never before, the demand for reform is emphasised by the great powers of Europe. This new constitution just promulgated in Constantinople contains no guaranties which the rest of Europe will not ultimately be obliged to secure from the populations of European Turkey. But if Islam must make the changes Europe demands, she must violate the Koran. Let adequate political reforms be perfected in Turkey, and Islamism is sure to unloosen her accursed, leprous grasp from the fair throat of the Bosphorus.

One of our most gifted missionaries and statesmen, Dr. Hamlin, has said lately, "Let Turkey stand, that Islam may fall." No doubt this opinion is a wise one from his point of view; and this morning even we, who are so little familiar with the politics of the Bosphorus, can understand, that, if all the reforms the recent conference of the great powers has asked for are carried, the Koran is a dead letter in Turkey. Dr. Hamlin seems to say that certain political changes are going forward in Turkey under the pressure of her own state necessities and of the demands of the great powers; that these changes cannot be carried through without violating in the boldest manner the political and religious provisions of the Koran; and that, therefore, if Turkey will carry these reforms through, she will undermine the authority of her own sacred book.

It seems probable, however, that Providence is to make shorter work with what Carlyle calls the unspeakable Turk than he would in any way make with himself under the pressure of the necessity for political reform. Is it not pretty clear that Gladstone's advice will ultimately be followed, and that Turkey as a Mohammedan empire will at least have no more armed support from Christian powers? If she must take care of herself, how long can she, who, in one of the fairest regions of the globe, is a treacherous bankrupt now, maintain her position in Europe, face to face with the increasingly angry protest of her own population and of Russia on the north, and of Austria, Germany, England, and France toward the setting sun? Constantinople and Cairo are held by Islam to-day only with faint grasp. Without these cities she will be driven back in her fearful sickness to her deserts. Only most slowly can she be healed there of her terribly poisoned blood. The days of the distinctively Mohammedan power in Europe are numbered.

Looking around the globe to-day, we see, therefore, an unbroken line of Christian influences in the near future, stretching from the Yosemite to the Sandwich Islands, to Australia, to Japan, to India, and past the Suez Canal, and thence to the Bosphorus, and thence to Germany, now possessing political and Protestant primacy in Europe, and so on to England, and then across that little brook we call the Atlantic, only two seconds wide now for electricity. There are no foreign lands.

In this year, America may say of her guests what was said of Portia's suitors:—

"The watery kingdom
Whose ambitious head threatens the face of heaven
Is no bar to stop the foreign spirits;
But they come as o'er a brook."

Merchant of Venice.

Christianity at this hour reads her Scriptures, and lifts up her anthems, in two hundred languages. One-half of the missionaries of the globe may be reached from Boston by telegraph in twenty-four hours. God is making commerce His missionary.

It is incontrovertible that it was predicted ages ago, that a chosen man called yonder out of Ur of the Chaldees should become a chosen family, and this a chosen nation, and that in this nation should appear a chosen Supreme Teacher of the race, and that He should found a chosen Church, and that to His chosen people, with zeal for good works, should ultimately be given all nations and the isles of the sea. In precisely this order world-history has unrolled itself, and is now unrolling. No man can deny this. No man can meditate adequately on this without blanched cheeks. What are the signs of the times which I have recounted on this festal morn, but added waves in this fathomlessly mysterious gulf-current? We know it began with the ripple we call Abraham. It is now almost as broad as the Atlantic itself. What Providence does, it from the first intends to do. We see what it has done. We know what it intended. It has caused this gulf-current to flow in one direction two thousand, three thousand, four thousand years. Good tidings, this gulf-current, if we float with it!—good tidings which are to be to all peoples! A Power not ourselves makes for righteousness. It has steadily caused the fittest to survive, and thus has executed a plan of choosing a peculiar people. The survival of the fittest will ultimately give the world to the fit. Are we, in our anxiety for the future, to believe that this law will alter soon? or to fear that He whose will the law expresses, and who never slumbers nor sleeps, will change His plan to-morrow, or the day after?

On this day of jubilee, let us gaze on this gulf-current, and take from it heart and hope, harmonious with the heart of Almighty God, out of which the gulf-current beats only as one pulse.

The difficulties that Christianity has now are chiefly in great cities. They are in the unfaithful members of highly civilised society. They are in that subtle and pernicious inactivity which undermines the nervous force of the world at its centres.

THE LECTURE.

De Wette, the great German theologian, who died in 1849, and who was called the Universal Doubter, said in his last work, published in 1848, that "the fact of the resurrection of Christ, although a darkness which cannot be dissipated rests on the way and manner of it, cannot itself be called into doubt" any more than the historical certainty of the assassination of Cæsar.[1] This is the passage over which Neander, the famous Church historian, shed tears when he read it. De Wette was a leader of the acutest school of rationalism in Germany in his day, and denied utterly that there are passages in the Old Testament Scriptures predicting the coming of our Lord. He was coupled by Strauss himself with Vater, as having placed on a solid foundation the mythical explanation of the Bible. Nevertheless, such is the cumulative force of the evidence of the resurrection as a fact in history, that De Wette, listening only to the latest voices of the most laborious, precise, and cold research, affirmed, face to face with the sneers of the rationalism which he led, that the fact itself, although we do not understand the way and manner of it, is incontrovertible.

---

[1] De Wette, Concluding Essay, appended to Historical Criticism of Evangelical History, p. 229.

I am to speak this morning, not of this fact, but of the way and manner of it. I know that the theme is fit to blanch the cheeks.

Before taking up this mystery of mysteries, however, let us, for a moment, glance at the logical value of De Wette's concession. It is a verdict reached unwillingly by long listening to all the public and secret words of history and philosophy—the guides which scepticism is so eager, and which religious science may well be yet more eager, to force upon the attention of the world.

I am accustomed to recite as a part of my private creed these propositions, based on De Wette's concession as to the fact of the resurrection :—

1. The intuitions of conscience prove the moral excellence of the biblical system.
2. The moral excellence of the biblical system proves that it is not inconsistent with the attributes of an infinitely perfect Being to give to that system a supernatural attestation.
3. If an historical attestation of this kind has been given to the biblical system, the existence of that attestation may be proved by the established scientific rules of historical criticism.
4. The established scientific rules of historical criticism, severely applied, demonstrate the fact of the resurrection.
5. The fact of the resurrection proves, not the Deity, but the Divine authority of our Lord, as a teacher sent into history with a supreme and divinely attested religious mission.
6. The Divine authority of our Lord proves the doctrines He attested.
7. Among these are His Deity, the Inspiration of the Scriptures, the necessity of the New Birth, the Atonement, Immortality, the Eternal Judgment.

It was my fortune once to put these propositions before the acutest intellect I have ever met in the field of theology, and to ask if they would bear the logical microscope. I remember, that, as I repeated them slowly, the majestic eyes of the listener were lifted from the earth to the horizon, and from the horizon to the infinite spaces of the Unseen Holy behind the azure. When at last I asked if De Wette's verdict did not contain in it all these conclusions, the unwavering reply was, " All, incontrovertibly. But De Wette's concession is the result of the conflicts of eighteen centuries of scholarship. Adhere to those propositions, for they have borne the tooth of time in the past, and will bear all the buffeting of acutest controversy in the future." Once in his garden at Halle-on-the-Saale, in an hour I shall long remember, I put those propositions before Professor Tholuck, with the same emphatic result.

It is on the way and the manner of the personal continuance of the soul after death that German philosophy now bends an intense, prolonged, reverent gaze. You will not suppose me to indorse everything which I put before you this morning as a part of the latest German philosophy. Nevertheless, I confess my sympathy with the whole trend of that magnificent body of thought which is represented by the Lotzes, the Helmholtzes, the Wundts, and the

Ulrici. Whoever is in accord with this school, which now leads the most intellectual and learned nation of our times, will find himself in most emphatic antagonism to the English materialistic school. This latter, however, has nothing to say that is new to Germany. Gentlemen here who have been accustomed to form their philosophical opinions from an English outlook, will, perhaps, allow me to ask them this morning for once, as an experiment, to occupy the German point of view. I do not request you to take the opinions of the Germans, though they have a far greater fame than the English for philosophical breadth and acumen; but will you not take their point of view long enough to understand that there are two philosophies in the world? If there is one represented by the Huxleys and Häckels, there is another opposed at all points to materialism, and represented by the Lotzes and Helmholtzes, and Wundts and Ulricis,—names which the future is far more likely to honour than those of any of their critics.

1. Lotze, Ulrici, Wundt, Helmholtz, Draper, Carpenter, and Beale, teach that the nervous mechanism in its influential arc is plainly so constructed that we must suppose it to be set in motion by an agent outside of it.

2. Every change must have an adequate cause.

3. *Only when involution is equal to evolution in the connection between cause and effect is the cause adequate to produce the effect.*

We all agree, and we talk smoothly, as to the authority of the tropically fruitful axiom, that every change must have an adequate cause. But what is an adequate cause? My definition, which I do not ask you to accept, is, *Such a cause as makes involution equal to evolution.* Sir William Thomson, speaking of the shrewd attempt of materialism to explain living tissues by infinitely complex molecular combinations of merely material particles, says it is for ever sure that we cannot get out of the combinations anything that we do not put into them; and that all science is against the idea that evolution can ever exceed, in the force or the design it exhibits, the involution which must go before the evolution. Involution before evolution is the fact on which to fasten attention, if we would be lifted out of materialism. *Let us be involutionists first, and evolutionists afterwards.* The astute attempt of Tyndall is to put into matter what he wishes to draw out of it. His whole effort is to introduce a new definition of matter. He would have us think of matter as a double-faced somewhat, having a material and spiritual side; and although, in attempting to do so, we necessarily fall into immeasurable self-contradiction, he is forced to undertake the support of even that, because he knows that evolution cannot be greater than involution. He would put into his theory, therefore, on the one side, that power and potency of all life which he wishes to take out on the other. It is the supreme law of philosophy that involution and evolution are an eternal equation. Materialism is marked by, perhaps, nothing more superficial than the attempt to avoid the force of that law in the explanation of living tissues. Even Tyndall,[1] after reasoning in favour of the theory which Professor Frey,

---

[1] Materialism and its Opponents, 1875.

the German histologist, says science has given up, that life is a kind of vital crystallisation, says inadvertently, with curious self-contradiction, that a living organism is " woven by a something not itself." Materialism astounds us by the assertion that physical and chemical forces are enough to explain the formation of living tissues; but no man has shown that in physical and chemical forces there can be an involution equal to the evolution we call organism and life. The evolution in man is intelligence, imagination, emotion, will, or all that we call the soul; and the involution, therefore, must have in it the equivalents of these qualities. For ever and for ever it will be true that you can find in living tissue, and take out of it, only what is put into it, visibly or invisibly.

4. The nature of what Aristotle called the animating principle, or the soul, is to be inductively inferred by an inflexible application of the principle that involution must equal evolution. In living tissues, as everywhere else, every change must have an adequate cause.

5. The co-ordination of tissues in a living organism must proceed from a sufficient cause, defined as one in which involution is equal to evolution, and which therefore must possess, not only intelligence, but permanence and unity in all the flux of the atoms of the body.

6. The unity of consciousness requires the same.

7. The persistence of the sense of personal identity requires the same.

The immense facts that each individual feels himself to be one, and that his identity through life is a certainty in spite of the flux of the particles of the body, are to be accounted for. It is enough to the acute German, born a metaphysician, to know that he has an ineradicable sense of personal identity, and that his consciousness is a unit, to cause him to repel the idea that all we call the soul is the result simply of an almost infinitely complex arrangement of atoms. Everywhere there is permanent unity in the plan of each organism that has life. All there is in the oak is woven after the fashion of the oak; all in the lion, after that of the lion; all in the man, after that of the man. We do know incontrovertibly that in each individual there is, from first to last, no deviation from the one plan on which the bioplasts weave. Now, that unity must be accounted for. It is a fact; it is tangible; it is visible.

If we have always before our speculative thought the ascertained activities of the bioplasts; if we behold them throwing out here and there their promontories, dividing and subdividing, and yet always weaving on a plan existing in the first stroke of their shuttles, and so carrying nerve around muscle, and forming here a vein and there an artery, here a tendon and there a hand, an ear, an eye, a brain,—we shall feel that all attempts to prove materialism by physiology are attempts to quench the noon under a bat's wing. Ulrici talks freely of much sand thrown in the eyes of our time by materialism; and so do Lotze and Helmholtz, and Wundt and Beale; and sometimes, in gusty days, I think there is a little of this dust even in this pellucid New-England air.

8. The nature of the animating principle has of late, in Germany, been very carefully inferred from the effects it produces.

It is the belief of many that science draws near to an explanation of some parts of the mystery in the connection of the soul with the body.

9. The late German philosophy holds the view that the soul must be conceived as a property or occupant of a fluid similar to the ether.

10. This fluid, however, does not, like the ether, consist of atoms.

Elaborate attempts to found the hope of existence after death on the scientific certainty that atoms cannot be destroyed have often been made; and an effort of this sort has lately appeared in the work of a New York authoress on "The Physical Basis of Immortality." She adopts Bain's philosophy, and talks of a material and a spiritual side in an atom; and she says that somewhere in the physical organism there is a soul-atom, and that this cannot be destroyed. This theory is German, only it is a little out of date, although Lotze once favoured it.[1] There are two competing theories,—that of the soul-atom and that of the soul-fluid. It is the doctrine of the non-atomic ether, or soul-fluid, which your Ulrici—whose German book, as you see, I read to pieces in a hundred miles in the railway train this morning—advocates. By the way, allow me to say that Ulrici's three volumes, entitled "Gott und der Mensch," published at Leipzig in 1874, are far more incisive than even his "Gott und die Natur" on all topics relating to living tissues and the connection between soul and body. Be sure to read the former work, especially the portion on the nervous system and the soul.[2]

It is Ulrici's view that the soul is the occupant of a non-atomic ether that fills the whole form, and lies behind the mysterious weaving of the tissues.

Who is Ulrici? Not a small philosopher, I assure you. Hermann Ulrici, professor of philosophy in the University of Halle, was born in Germany in 1806. He studied law and afterwards physical science in the stern manner of the German universities, and then gave himself to literature and philosophy. He has written an elaborate work on æsthetics; and his criticisms on Shakspeare are the best, except those of Gervinus. Everywhere in Germany he is recognised as authorised to speak on the nerves and the soul from the point of view of a specialist; and his is, perhaps, the highest name in Germany, after that of Lotze, in all philosophy connected with the relations between mind and matter.

11. This non-atomic fluid is absolutely continuous with itself.

12. Its chief centre of force is in the brain.

13. But it extends outward from that centre, and permeates the whole atomic structure of the body.

Have you ever, my friends, floated in thought above the green and steel-grey seas of the globe, and called vividly before your imagination the contrast between the dark depths and the sunny

---

[1] For Lotze's present views, see Mikrokosmus, Drittes Buch, Zweites Kapitel, Von dem Sitze der Seele, Allegegenwart der Seele im Körper.
[2] Vol. i. pp. 161-225; see, also, Ueberweg's History of Philosophy, vol. ii. p. 303.

surfaces of the oceans? The upper portions of every ocean are permeated by the sunbeams; but, as we descend in the Atlantic or Pacific, we come to obscurity; and, in the lowest search of the sea, there is darkness. Just so in the connection of the soul with the body. There is a sunny sea, an obscure sea, and a dark sea. A portion of the operations of the immaterial principle in us we are vividly cognisant of through consciousness. A few of the activities of our physical organisation we are conscious of obscurely; most of them, however, and all this weaving of tissues, go on wholly below consciousness. There seem to be mental operations that proceed in the darkness of the mental Atlantic. Some go on obscurely in a region of partial illumination. But intellect, will, emotion, belong to those sunlit waves where consciousness fills the billows at the surface of the mental ocean with iridescence. You will readily admit that consciousness does not make us aware of all the activities of the immaterial principle. That unit which we call the soul is not cognisant of all its own operations as it is conscious of memory or of an act of reason. Many things which the immaterial principle in man does, it performs in the dark depths, where no man's consciousness comes, and yet God is there.

14. The soul, as an occupant of this ethereal enswathement, operates in part unconsciously, and in part consciously.

15. It co-operates with the vital force.

16. It is not identical with that force.

In order to explain living tissues, it is not necessary to assume the existence of what is called vital force; but it is necessary to assume the existence of an immaterial principle. Hermann Lotze takes great pains, and Ulrici does, to show that the immaterial principle is not necessarily to be thought of as identical with what has been called the vital force. That which moves these bioplasts, and causes them to build on a plan kept in view from the first, and maintained as a unit to the last, we say must be an adequate cause of these motions; and that is not the vital force simply, although it may be the vital force with this other psychical force behind it; and yet the two are always to be carefully distinguished from each other.

17. The soul has a different type for each different organism.

As it were folded up, it exists, of course, in the embryonic germ of each organism,—oak, lion, eagle, or man.

18. It is the morphological agent which weaves all living tissues. It spins nerves. It weaves the muscles, the tendons, the eye, the brain. It arranges each part in harmony with all the other parts of the organism.

19. When it rises to the state of consciousness, it produces the phenomena known as thought, imagination, emotion, and will.

20. So far forth as the ethereal enswathement of the soul is nonatomic, it is immaterial.

It is the business of the Boston Monday Lectureship to keep before this audience, so many members of which know more than the lecturer, the very latest speculations, if they lead to anything strategic. You will allow me to say that as wise men as Martineau and Ulrici and Beale and Lotze and Helmholtz do not sneer at the idea that the universe may have in it three things, and not merely two. Matter and mind, we have commonly said, include everything; but some are whispering, " Perhaps there is an invisible middle somewhat, for which we have no name, but which is remotely like the ether." Is it material? It is not atomic; and matter is. Now, Ulrici so far adopts this idea as to affirm explicitly that the ethereal enswathement of the soul must be non-atomic, and so far not like matter. He thinks that the atomic constitution of this enswathement would be absolutely inconsistent with the fact of the unity of consciousness. He holds, that, if the soul-fluid be made up of atoms, there is no proof that it is not in flux with the flux of the particles of the body. But the persistence of our sense of individuality is proof that there is no such flux in the substance in which mental qualities inhere. We know that there are in us certain mental attributes, and that every attribute must have a substratum; and in the substratum in which anything permanent, like the sense of identity, inheres, there must be no flux, but permanence. Therefore, following the clew that every change must have an adequate cause, Ulrici holds that the enswathement of the soul, this ethereal body, is non-atomic, and not in flux.

Just as the summer lightning blazes through the cloud, so the soul blazes through that spiritual body which is finer than nervous tissue, finer than electricity. When the egg begins to quicken, the life is the chief thing in it, and that life belongs to a certain somewhat, an ethereal form of matter that connects it with all this dead world around. The soul inhering in that spiritual body takes to itself clothing, and builds the visible matter upon the invisible. According to the law of the invisible matter, according to its power to take large or small space as its exigencies require, it grows, for a season, larger and larger, until the soul in it has taken clothing to itself out of this visible world. We appear here as ghosts appear in the night. Carlyle says we are all ghosts; we appear, we disappear; we come forth from the invisible, we go into the invisible. These are facts; but Germany begins to speculate as to the adequate causes of our being woven as we are, and says that, behind all the weaving of our tissues, there must be this ethereal body. Why does she say that? Germany commonly has a reason for her positions.

There is Niagara. You see a rainbow drawn across the surface of the cataract. The rainbow does not move. The water moves. What is the cause of the rainbow? The water, you say. No! Germany replies; the rainbow never moves. If the water were the chief cause of the rainbow, the rainbow would move; for you must have in the fountain what you have in the source. The occasion of the rainbow is in the water; the cause is in the sun. That is not in flux. Your rainbow is not in motion, either. Now, the plan of man's organism does not change from the first quickening of the egg until the man drops into the grave. It is one thing, just as that rainbow is one thing. Our sense of identity persists. Nevertheless, all the particles in the body are changing as the drops in Niagara are. The cause of our sense of personal identity must be something that is not in perpetual change. Your fountain cannot rise higher than your source. The plan of your mechanism does not change, and so the source of that plan does not change. We know that every coarser physical particle does change. There is nothing in my hand that was there seven years ago, I suppose, except the plan of the material. The particles have all been changed; but the plan is just the same. That plan which does not change implies the existence in man of a substance which does not change, and, although that substance is invisible, science thinks it is there because it sees effects which can be explained only upon that supposition.

We know that the rainbow is not in flux, and so we know there is something behind it which causes it to persist in one form. As the plan of your eagle, your lion, your man, your oak, is steadily adhered to from first to last, we say that plan belongs to something that is not in flux, that came in when the plan threw its first shuttle, and goes out unimpaired, even after the shuttle ceases to move. That invisible somewhat some scholars in Germany call a spiritual body.

21. This non-atomic ethereal enswathement of the soul is conceivably separable from the body.

How shall I proceed, gentlemen, when thoughts crowd upon us here and now that soon will seem too sacred even for the hushed chambers from which you and I must pass hence, each alone? Who has treated death inductively? What do the dying see? What do they hear? What do they fear, and what do they hope? I am asking of you only loyalty to the self-evident truth, that every change must have an adequate cause. The Ariadne clew has now brought us mercilessly up to the certainty that the adequate cause of all this weaving of living tissues must be something having unity; something not in flux with the constant changes of the particles of the body; something that is as steady as the rainbow drawn across the east, while all the drops of rain are rapidly changing their position.

It is not every untrained or trained mind that is able to follow even this axiomatic Ariadne clew through all this labyrinth of philosophy. Sometimes I think that philosophers are to be divided into classes like generals, according to their capacity to manage intricate problems. There are generals that can command ten thousand men; but Napoleon said, there are only a few who can command five hundred thousand. There are intricacies in philosophy which it takes a Lotze or an Ulrici, a Kant or a Hamilton, a Helmholtz or a Beale, to walk through without bewilderment. Adhere to the writers who are clear. Many a general on the field of philosophy can take care of ten thousand; but only now and then one can manage five hundred thousand men.

If you come to the conclusion that there is an invisible, non-atomic, ethereal enswathement, which the soul fills, and through which it flashes more rapidly than electricity through any cloud, you must remember that the majestic authority for that statement is simply the axiom that every change must have an adequate cause. This is cool precision; this is exact research on the edge of the tomb. Professor Beale says in so many words, "that the force which weaves these tissues must be separable from the body;" for it very plainly is not the result of the action of physical agents. Ulrici shows, especially in a magnificent passage on immortality,[1] that all the latest results of physiological research go to show that immortality is probable.

You say that, unless we can prove the existence of something for the substratum of mind, we may be doubtful about the persistency of memory after death; but what if this non-atomic, ethereal body goes out of the physical form at death? In that case, what materialist will be acute enough to show that memory does not go out also? You affirm, that, without matter, there can be no activity of the mind; and that, although the mind may exist without matter, it cannot express itself. You say, that unless certain, I had almost said material, records remain in possession of the soul when it is out of the body, there must be oblivion of all that occurred in this life. But how are you to meet the newest form of science, which gives the soul a non-atomic enswathement as the page on which to write its records? That page is never torn up. The acutest philosophy is now pondering what the possibilities of this non-atomic, ethereal body are when separated from the fleshy body; and the opinion of Germany is coming to be very emphatic, that all that materialists have said about our memory ending when our physical bodies are dissolved, and about there being no possibility of the activity of the soul in separation from the physical body, is simply lack of education. There is high authority and great unanimity on the propositions I am now defending; and although I do not pledge myself always to defend every one of these theses, yet I must do so in the present state of knowledge and in the name of a Gulf-current of speculation which is twenty-five years old, and has a very victorious aspect as we look backward to the time when the microscope began its revelations.

22. It becomes clear, therefore, that, even in that state of existence which succeeds death, the soul may have a spiritual body.

What! you are preaching to us the book called the Holy Word? Yes, I am; and here is a page of it [with a hand on coloured diagrams of living tissues]. A spiritual body! That is a phrase we did not expect to hear in the name of science. It is the latest whisper of science, and ages ago it was a word of revelation.

23. The existence of that body preserves the memories acquired during life in the flesh.

24. If this ethereal, non-atomic enswathement of the soul be interpreted to mean what the Scriptures mean by a spiritual body in distinction from a natural body, there is entire harmony between the latest whisper of science and the inspired doctrine of the resurrection.

What if I should dissect a human body here? I might have a man made up of a skeleton; then I could have a human form made up of muscle. If I should take out the arteries, I should have another human form; and just so with the veins, and so with the nerves.

---

[1] Gott und der Mensch, vol. i. pp. 222-225.

Were they all taken out and held up here in their natural condition, they would have a human form, would they not? Very well; now, which form is the man? Which is the most important? But behind the nerves are those bioplasts. If I could take out those bioplasts that wove the nerves, and hold them up here by the side of the nerves, all in their natural position, they would have a human form, would they not? And which is the man? Your muscles are more important than your bones; your arteries, than your muscles; your nerves, than your arteries; and your bioplasts, that wove your nerves, are more important than your nerves. But you do not reach the last analysis here; for, if you unravel a man completely, there is something behind those bioplasts. There are many things we cannot see that we know exist. I know there is in my body a nervous influence that plays up and down my nerves like electricity on the telegraphic wires. I never saw it; I have felt it. Suppose that I could take that out. Suppose that just there is my man made up of nerves, and just yonder my man made up of red bioplasts; and that I have right here what I call the nervous influence separated entirely from flesh. You would not see it, would you? But would not this be a man very much more than that? or that? What if death thus dissolves the innermost from the outermost? We absolutely know that that nervous influence is there. We know, also, that there is something behind the action of these bioplasts. If I could take out this, which is a still finer thing than what we call nervous influence, and could have it held up here, I do not know but that it would be ethereal enough to go into heaven, for the Bible itself speaks of a spiritual body. You know it is there, this nervous influence. You know it is there, this power behind the bioplasts. When the Bible speaks of a spiritual body, it does not imply that the soul is material; it does not teach materialism at all; it simply implies that the soul has a glorified enswathement, which will accompany it in the next world. I believe that it is a distinct ¦biblical doctrine that there is a spiritual body as there is a natural body, and that the former has extraordinary powers. It is a body which apparently makes nothing of passing through what we call ordinary matter. Our Lord had that body after His resurrection. He appeared suddenly in the midst of His disciples, although the doors were shut. He had on Him the scars that were not washed out, and that in heaven had not grown out. I tread here upon the edge of immortal mysteries; but the great proposition I wish to emphasise is, that science, in the name of the microscope and the scalpel, begins to whisper what revelation ages ago uttered in thunders, that there is a spiritual body with glorious capacities.

This is a sad world if death is a leap in the dark. But, gentlemen, we are following haughty axiomatic certainty. In clear and cool precision, science comes to the idea of a spiritual body. We must not forget that this conclusion is proclaimed in the name of philosophy of the severest sort. The verdict is scientific; it happens also to be biblical. Is it the worse for that? It is more and more evident, as the training of the world advances, that everything fundamentally biblical is scientific, and that everything fundamentally scientific is biblical.

In every leaf on the summer boughs there is a network which may be dissolved out of the verdant portion, and yet retain as a ghost the shape which it gave the leaf from which it came. In every human form, growing as a leaf on the tree Igdrasil, we know that network lies within network. Each web of organs, if taken separately, would have a form like that of man. There might be placed by itself the muscular portion of the human form, or the osseous portion, or the veins, or the arteries, and each would show the human shape. If the nerves could be dissolved out, and held up here, they would be a white form, coincident everywhere with the mysterious, human, physical outline. But the invisible nervous force is more ethereal than this ghost of nerves. The fluid in which the nervous waves occur is finer than the nervous filaments. What if it could be separated from its environment, and held up here? It could not be seen; it could not be touched. The hand might be passed through it; the eyes of men in their present state would detect no trace of it; but it would be there.

Your Ulricis, your Lotzes, your Beales, adhere unflinchingly to the scientific method. The self-evident axiom, that every change must have an adequate cause, requires us to hold that there exists behind the nerves a non-atomic, ethereal enswathement for the soul, which death dissolves out from all complex contact with mere flesh, and which death, thus unfettering without disembodying, leaves free before God for all the development with which God can inspire it.

> " Then long Eternity shall greet our bliss
>   With an individual kiss,
>   And joy shall overtake us as a flood,
>   When everything that is sincerely good
>   And perfectly divine,
>   With Truth and Peace and Love, shall ever shine
>   About the supreme throne
>   Of Him to whose happy-making sight, alone,
>   When once our heavenly-guided souls shall climb
>   Then, all this earthly grossness quit,
>   Attired in stars we shall for ever sit,
>   Triumphing over Death and Chance and thee, O Time!"
>
> MILTON.

THE END.

INDEX.

―――o―――

ABIOGENESIS. See *Spontaneous Generation*.
Abraham. Call of, 139.
Action of matter in the mineral world and in living tissues, 70.
Adaptation. Darwin on an instance of, 30.
Æschylus on retribution, 64.
Agamo Genesis, 55.
Agassiz on Darwinianism, 31.
―― on the immortality of instinct, 97.
Agnosticism, a transient stage in culture, 135.
―― its varieties, 5.
―― the ultimate form of scepticism, 106.
Ariadne clue of scientific investigation, the, 9, 47, 146.
Aristotle and Bacon, 18, 19.
―― on axiomatic truths, 102.
―― on the animating principle, 81.
Atheism. Frederick the Great on, 49.
―― Lotze on, 49.
―― Tyndall on, 9.
―― Wurtz on, 12.
Atlantic. The : a brook, 138.
Atoms are in themselves incapable of building up organisms, 68, 69.
―― Indestructibility of, 143.
―― marvels attributed to their fortuitous concourse, 67.
Automata. Distinction between the animals and, 81, 97.
―― In what sense men are, 81, 83.
Automatic action in creatures from which the brain has been removed, 115, 116.
―― and free activities in man, 82.
―― and influential nerves, 115-125.
―― and nervous influential actions. Contrast between, 115, 116.
―― nervous arcs, 78, 80, 94, 95, 96, 118.
Australia, its future, 137.
Awaking from sleep, 81.
Axiomatic truths. Aristotle on, 102.
―― furnish a safe standing ground, 30.
―― the foundation of all science, 119.
―― the foundation of metaphysics, 106.
―― their helpfulness, 101, 102, 106.

BACON—the services he rendered to philosophy, 19.
Bain, his admission concerning the separation between the organic and the inorganic, 104.

Bain, his confusion of "union" with close succession, 105.
―― his materialism, 101-112.
―― his "Mind and Body," 52, 104.
―― his theory of the spontaneity of vital actions, 69.
―― on the spontaneity of vital actions, 69.
Barry on the cell theory, 50.
Bastian's works, 51.
Bathybius Häckel and Beale on, 3.
―― Mirvart on, 34.
―― Huxley's varying accounts of, 1-4.
―― Strauss on, 31.
Beale, his method of microscopic research, 38, 60.
―― his "Protoplasm, or Matter and Life," 36, 113.
―― on the arrangement of matter, 113.
―― on the microscope, 51.
―― Testimony of Drysdale concerning, 52.
Bible and science. Harmony between the, 148, 149.
Biblical teaching concerning the spiritual body, 148.
Biology. Inverse solution of problems in, 84.
―― its interest as a study, 70.
―― Standard works on, 113, 114.
―― the three most important distinctions in, 8, 117.
Bioplasm a better term than protoplasm, 39.
―― an essential element in all living organisms, 37.
―― Facts concerning, 58-68.
―― is apparently structureless, 46.
―― is directed by life, 65.
―― is the same thing in every tissue, 44, 62, 72.
―― its change of form, 62.
―― its marvellous capacities, 47.
―― meaning of the term, 39.
―― See also *Protoplasm*.
―― the original, 67.
―― the physical basis of life, 36.
―― variety of its activities, 109.
Bioplasmic theory. Central truth of the, 50.
Bioplasts are apparently alike in all living forms, 42, 72.
―― are the units of growth, 39.
―― Carpenter on, 41.
―― Confession of Huxley and Häckel concerning, 43.

INDEX.

Bioplasts described, 44.
—— Growth of, 61.
—— how they are built up into living organisms, 144.
—— indicate intelligence somewhere, 41, 73.
—— marvellousness of their activities and products, 142.
—— pervade all living structures, 38.
—— their movements, 71.
—— their origin and action described, 39, 41, 43.
Birks on fatalism and evolution, 102.
Body. The : its changes and continuity, 12, 13, 82, 116, 145.
—— its relation to the soul, 12, 65, 78, 91, 93-96, 109-111, 144.
—— its wonderfulness and sacredness, 111, 112.
—— The ineffaceable contrast between the soul and, 78.
Border-land between the physical and the spiritual, 47.
Boston and Edinburgh, 126.
Brainless animals. Experiments with, 81.
Brain. The : changes in the cells of, 121.
—— Effects of electrical stimulation of, 93.
—— —— mutilation of, 94, 95, 108, 121.
—— —— sensation on, 122.
—— Facts concerning, 118, 119.
—— Ferrier on, 92, 95, 96, 107.
—— Functions of the hemisphere of, 121.
—— —— Lower ganglia of, 121.
—— in man and in the ape, 22, 23, 31, 33.
—— is in itself inert, 85, 96.
—— is the chief centre of force, 143.
—— is the organ of the mind, 121.
—— is woven by bioplasts, 78.
—— Localisation of functions in, 93, 94, 96.
—— Physiological and psychological activities of, 96, 107.
—— the meeting-place of the influential arcs, 84.
Bremer. Fredrika : her intercourse with Emerson, 131.
Brown. Alexander : on the cell theory, 50.
Bryant on the immortality of the soul, 136.
Butler, his "Analogy," 15.
—— on conscience, 134.
Butterfly. The : its history, 110.

CABANIS, his theory of the soul, 26.
Carlyle on Darwinianism, 63.
—— on newspapers, 26.
—— on the soul, 111.
Carmine, its use in microscopic research, 38, 60.
Carpenter, his "Mental Physiology," 11.
—— his theory of the location of consciousness, 118.
Cause and effect, 107.
Cause of motion, 74.
Causation. Belief in, 30.
—— is the will of God, 25.
Cellated nervous arcs, 79.
Cellated theory. History of the, 36, 49, 50.
—— is now discarded, 39.
—— Schwann on the, 70.
—— Tyson on the, 70.

Cells examined under the microscope, 53, 54.
Cell walls are formed matter, 38, 39.
Cerebration. Unconscious, 11.
Challenger. The—result of its explorations, 2, 3, 34.
Chemical and physical forces, 65.
Chemistry fails to explain the transformation of pabulum into living matter, 38-41.
—— its inability to produce life, 68.
Chloroform, its value, 98.
Christ : His appearance after His resurrection, 148.
—— The resurrection of, 139, 140.
—— prophecies of His coming, 139.
Christianity, extent of its influence, 137-139.
—— its relation to science, 87.
Christian theism, 128.
Chrysalid and butterfly, 110.
Church. The : its influence on civilisation, 75.
Civil Service Reform, need of, in America, 45, 101.
Clearness of thought. Importance of, 11, 40, 122.
Colleges, what they teach, 113.
Colouring living tissues. Beale's method of, 38.
Conception. The miraculous, 55.
Concessions of evolutionists, 18-34, 67-69.
Conscience, its power, 128.
—— its prophetic action, 134, 135.
—— its relation to the influential arcs, 80, 81.
—— trustworthiness of its testimony, 135.
Consciousness. Carpenter's theory of the location of, 118.
—— Phenomena of, 68.
—— Testimony of, 124.
—— The unity of, 73, 74, 111, 142.
Conversation meetings, 127.
Co-operation, 75.
Co-ordination of parts in living tissues, 47, 62, 142.
Creation. Biblical account of, 5.
—— Design in, 15.
—— is the act of God, 25.
Creative personal First Cause. Hypothesis of a, 13, 16, 18, 25.
Crystals. Formation of, 40.
Culture. Four stages in, 135, 136.

DANA on the origin of man, 22.
Darwin, his agreement with Butler as to the term "natural," 15.
Darwin, his theory of natural selection, 4, 6, 15, 20, 21, 23, 24.
—— his theory of the survival of the fittest, 23.
—— is no longer a good Darwinian, 24, 29.
—— relation of his theory to religion, 15, 16.
Darwinianism. Carlyle on, 63.
—— Owen on, 40.
Death : does it end all ? 64, 99, 102-112.
—— Premonitions of, 89.
—— what it accomplishes, 148.
—— why men dread it, 134.
Deduction and induction. Difference between, 33.
Deductive conclusions are sometimes lawful, 33.
Definitions. Importance of, 65, 103.

INDEX.

Democratic ages. Perils of, 25.
Design in creation, 15, 16, 32, 98.
—— the arguments from, 15.
Desires. Natural : what they imply, 132.
Development. Doctrine of. See *Evolution*.
Divine nature. Immanency and Transcendency of the, 129.
—— will. The : its connection with matter, 125.
Doges of Venice, 46.
Doubt, how it is to be vanquished, 101.
Draper's "Conflict between Science and Religion," 85.
—— his declaration concerning the soul, 86.
—— "Intellectual Development of Europe," 12.
Drysdale on the protoplasmic theory, 52, 114.
—— on the "stimulus" of matter, 69.
Dying. Visions of the, 146.

EAR. The, 84, 109.
Earth. Age of the, 27, 28.
Edinburgh. Mr. Moody's work in, 126, 128.
Electrical currents, their effect on the brain, 94, 193.
—— stimulation of the nerves, 78, 79.
—— —— automatic arcs, 95.
Electric eels, 37.
Embryos, 43, 44.
Emerson, disorder of his works, 129.
—— his farewell letter to his parish, 131.
—— his guiding principle, 129, 130.
—— his inconsistencies, 129, 132.
—— his individualism, 129, 130.
—— his intercourse with Fredrika Bremer, 131.
—— his mission, 129.
—— his pantheism, 129, 130.
—— his views on immortality, 128-136.
Emotion, its relation to the automatic arcs, 82.
England and the English. Characteristics of, 74.
—— and Russia, 137.
—— her Eastern possessions and influence, 137.
Evolution and involution are a fixed equation, 68, 107.
—— Doctrine of : Birks on the, 102.
—— —— Carlyle on the, 63.
—— —— Case against the materialistic, 53.
—— —— Fatal flaws in Huxley's, 19, 20, 31.
—— —— is not necessarily antagonistic to religion, 15, 32.
—— —— its demands in regard to time, 26-28.
—— —— its relation to biology, 1.
—— —— Jevons on the, 32.
—— —— Kingsley on the, 16.
—— —— Lotze on the, 49.
—— —— Missing link in the, 22, 31.
—— —— Objections to, 26, 27, 31.
—— —— Theistic forms of the, 4.
—— —— the rock on which the radical form of it is wrecked, 43, 68.
—— —— Tyndall on the, 10.
—— —— Various forms of the, 41, 53.
Evolutionists. Concessions of, 18, 34.
—— Varying schools of, 4, 30, 33, 53.

Existence after death. Expectation of, 133, 134.
Extemporaneous speech, 82, 83.
External conditions, their influence, 30.
Eye of the trilobite, 29.

FACTS in science. Alleged : how are they to be tested, 51.
Ferrier on the brain, 93, 107, 121.
Force. Difference between physical and vital, 67.
—— defined, 9, 92.
—— its spiritual origin, 16, 92.
—— Persistence of, 68.
—— Relations of chemical and vital, 122, 123.
Forces and motions are not the same, 78.
—— Chemical and physical, 65.
Forecast, evidences of, in nature, 42, 73, 113.
Form in physical organism. Cause of, 81.
Fossils. Human, 22.
Freedom. Elements of popular, 75.
Frey, his works on histology, 114.
—— on microscopic technology, 51, 69.
—— on the "cellular theory," 69.
—— on vital transformation, 70.
Frog. Huxley's experiments with the headless, 81.
Frogs, their automatic actions, 115.
Frontal development and intellectual power, 96.

GALVANIC currents, their effects on the influential arcs, 94, 95, 96.
—— nervous arcs, 78, 94, 95.
Ganglia of nerves, 79.
Gaseous state. Life incompatible with the, 26.
Generation. Non-sexual, 55.
Genesis and geology, 5.
Geological time, 27.
Geology. Text-books on, 36.
Germs. Primordial, 7.
Goethe on the immortality of the soul, 57.
Goodsir on the cell theory, 50.
Gravitation. Discussions concerning, 10.
Grecian discussions concerning the soul, 64.
Gyges' ring, 76, 77.

HABIT, its relation to the automatic arcs, 82, 83.
Häckel. Accounts of, 27, 40.
—— his fatalism, 102.
—— on the protoplasmic theory, 50.
—— on the transition from inorganic to organic matter, 3.
Hamilton, Sir W., his interest in mesmerism, 124.
—— his method of discussion, 49.
—— on mind and matter, 14.
Harp and the harper symbolical of body and soul, 67, 89, 109.
Harvard University. Teaching at, 86.
Henle on the cell theory, 50.
Herschel, Sir John, on the soul, 84.
Historical retrospects, their value, 49.
Human fossils, 22.
Huxley, defects and inconsistencies of his "New York Lectures," 3.
—— German estimate of, 103.

INDEX. 153

Huxley, his admission concerning the sterility of hybrids, 21, 22.
—— his admission concerning the will, 82, 115.
—— his concessions as to spontaneous generation, 19.
—— his determined agnosticism, 20.
—— his experiments with a headless frog, 81.
—— his materialism and fatalism, 102, 103.
—— incompatibility of his theories with those of Tyndall, 20.
—— on the differences between man and the apes, 31.
—— on " Molecular Machinery," 103.
—— on the physical basis of life, 4.
—— on protoplasm, 56.
Hybrids, sterility of, 21, 34.
Hylozoism, 9, 14.

IDEALISM, 73.
Identity. Personal, 12, 13, 110. 111, 142, 145.
Immortality of the lower animals, 97.
Immortality of the soul, Arguments for the—
1. Life is the cause, not the result, of organisation, 44, 67.
2. The soul is indestructible, 56.
3. Its relation to the body is that of a rower to a boat, 44.
4. The soul is independent of, and external to, the body, 81, 84, 85, 96, 109, 124.
5. The particles of matter in our body are continually changing, and yet the sense of personal identity is continually preserved, 110, 111.
6. In the metamorphoses of insects, though the particles of the body are changed, and therefore entirely altered, yet their life and identity are preserved, 110.
7. Belief in the immortality of the soul is instructive and well-nigh universal, 133, 134.
8. The demand of the moral law for " perfection " contains in it a postulate of immortality, 134.
9. Conscience prophesies that there awaits us a future of righteous retribution, 135.
—— Bryant on the, 136.
—— Emerson's views on, 128, 136.
—— is almost universally expected and desired, 133.
—— is scientifically probable, 147.
—— Kant on the, 134.
—— Mahomet on the, 75.
—— suggested by the microscope, 43.
—— The physical basis of, 43.
—— Transcendentalism and the, 132.
Induction and deduction. Difference between, 33.
Inductive method. The, 14.
Inertia defined, 9, 68.
—— the invariable characteristic of matter, 68, 69.
Influential arcs, 78-81, 83.
Inherited tendencies, 83.
Instinct, is it immortal? 97.
—— its relation to the automatic arcs, 80.

Instinct, its relation to the nervous mechanism, 133.
Instincts. Organic : what they imply, 132-136.
Intellect, its relation to the influential arcs, 81.
—— The seat of, 94.
Intellectual principle. Draper on the, 86.
Internal activity not always dependent on external irritation, 81.
Internal senses. Nerves of the, 118.
International scientific series. The, 52.
Intuitions are a test of verity, 73.
—— Testimony of the, 14.
Inquiry meetings, 127.
Involution and evolution, 68, 141, 142.

JACKSON'S political principle, 101.
Japan, its future, 137.
Jevons on the doctrine of evolution, 32.

KANT on the immortality of the soul, 134.
—— necessity of a cause external to nature, 20.
Knowledge. Limits of, 135.

LAW. Natural : its uniformity, 18, 21
—— Relation of God to, 18.
—— Tennyson on, 17.
Laws of nature. Mill on the, 113.
Leydig on the cell theory, 50.
Life and mechanism are two things, 108.
—— can neither be produced or explained t by chemistry, 40.
—— compared to a rower in a boat, 66, 67, 89, 96.
—— Definition of. 65.
—— Huxley on the physical basis of, 2, 4.
—— is not the result of mechanism, 56.
—— —— organisation, 43, 44, 67.
—— is the formative power in nature, 71.
—— its fundamental phenomena, 50.
—— its origin, 43, 67.
—— Lowell on the basis of, 93.
—— may exist before or after organisation, 71.
—— origin of : Häckel on the, 34.
—— —— Sir W. Thomson's theory, 20.
—— —— Tyndall on the, 9.
—— or mechanism—which ? 57-63.
—— The physical basis of, 36, 54.
—— Tennyson on the mysteries of, 44.
Lincoln, President, his early religious difficulties, 55.
Living organisms consist of three parts, 37.
Living tissues. Agreement of Huxley and Beale concerning, 58.
—— Lotze, Beale, and Huxley on, 46-56.
—— Sir W. Thompson on the origin of, 141.
—— the interest that attaches to investigation of, 58.
Logical laws. The first of all, 104.
Lotze. Accounts of, 48, 114.
—— his supreme arguments against materialism, 73.
—— the great maxim of his philosophy, 108.
Lowell on the basis of life, 93.

MACAULAY'S essay on Lord Bacon, 18
Madonna di San Sisto. 36.
Magnets and organic life, 40

L

INDEX.

Mahommedanism—its future, 137, 138.
Man. Carlyle on, 63.
—— consists of body, soul, and spirit, 147.
—— has not descended from the Ape, 22, 23, 31, 32, 33.
—— his complete nervous structure, 80, 81, 82.
—— his origin, 30, 33, 34.
—— is the true Sheckinah, 111.
—— The hairlessness of, 23.
Materialism. Forms of, 7, 8.
—— Illustration of the logic of, 123.
—— is discredited by the latest science, 62, 108.
—— its decline in Germany, 8, 74.
—— its failures, 141.
—— its failure to explain the collocation of parts, 13.
—— its fundamental absurdity, 105.
—— its self-contradictions, 106.
—— its stupidity, 108.
—— Lotze's supreme argument against, 73
—— the rock on which it is wrecked, 68, 105, 113.
—— what it teaches, 35, 36, 40.
Materialistic doctrine of evolution. Case against the, 53.
Matter and mind, 10–14, 104, 117.
—— Arrangement of, 72.
—— Bain and Tyndall's definition of, 52, 103, 104.
—— Beale on the arrangement and control of, 113.
—— Divers actions of, 70.
—— is an effluence of the Divine nature, 125.
—— not a double-faced unity, 12–14, 103–108.
—— Properties of, 9, 13.
Mechanism, its mission in the universe,108.
Memory is not a result of our physical organisation, 147.
—— its relation to the nervous mechanism, 79.
—— Physical basis of, 12.
Mental operations performed when half the brain is removed, 95.
—— are often obscure, 144.
Mesmeric force, 124.
Metaphysics and physiology—their harmony, 117.
—— defined, 106.
Meyer's definition of force, 92.
Microscope. The: and materialism, 35, 44.
—— Beale on, 59.
—— its suggestions of immortality, 43.
—— Power of, 46.
Microscopic research. Results of, 36, 44, 53–55, 108.
Mill. John Stuart: on the arguments from design, 15.
Mind. Connection between matter and, 10–12, 104, 117.
—— Distinction between matter and, 14, 16.
—— its immateriality, 14.
—— mystery of its influence on matter, 124.
—— Properties of, 13.
Ministers, their capacity and their right to take part in philosophical discussions, 47, 48, 119.
Ministry. Training for the, 121.
Miracle defined, 15.

Miracles. Scientific proof of the possibility of, 34.
Miraculous conception. The, 55.
Mivart on Darwinianism, 24, 30.
Molecular law, 32.
—— machinery, 40, 62, 72, 103.
—— motions in the nervous system, 124.
Molecules. Professor Maxwell on the existence and qualities of, 1.
Moneres. Häckel on the, 2.
Moody, Mr., effect of his preaching, 128.
—— his work in Edinburgh, 126–128.
—— the secret of his usefulness, 127.
Moral law. The: demands perfection, 134.
Motion and force not the same, 78.
Motion. Cause of, 92.
Mound boulders of the Mississippi, 66.
Muscular fibre. Formation of, 41.
—— motion produced by electric action on the brain, 94.
Mysteries are not self-contradictions, 124.

Nägeli on the cell theory, 50.
Natural, definition of the term, 14, 15.
Natural law. God's relation to, 128.
Natural selection. Darwin's theory of, 20, 21, 23, 24.
—— theology, its place in college studies, 119.
—— theology. Spencer on, 5.
Nature of things. Socrates on the, 101.
Nature. The animating principle of,142, 143.
—— the unity of, 27.
Nebular hypothesis. The, 7, 11, 20, 32.
Nerve centres, 79.
Nerves. The: and the soul, 76–88.
—— Automatic and influential, 115–125.
—— end in loops, 51.
—— Formation of, 41.
—— The mechanism of, 78, 86.
—— wondrousness of their formation, 62.
Nervous activities, automatic and influential, 117, 118.
Nervous fibres, 78.
—— influence. Transmission of, 79.
New lands, where they are to be discovered, 58.
Newspapers, 26.

Open secret. The, 15.
Organ. The invisible player on the, 61.
Organic structures, their mysteries, 47.
Organic. The: its separation from the inorganic, 70.
Organisms are all formed according to a plan, 42.
—— cannot be produced by atoms, 68, 69.
—— Living: their elements, 37, 38.
—— unity of their plan, 142, 146.
Organisation and life. Relation of, 67, 71, 72, 73.
—— is a product of life, 71, 73.
—— is not the cause of life, 43, 44, 67.
Origin of species, various hypotheses, 4.
Osborne, Lord S. S., on the cell theory, 50.
Over-soul, Emerson's loyalty to the, 129, 130.
Owen on Darwinianism, 40.

Paradise, conception of, by Agassiz, 97.

INDEX.

Pabulum, 38.
Paganism, significance of its efforts to propitiate supreme powers, 135.
Pantheism, a half truth, 128.
―― Emerson's, 129, 130.
―― summary of its teachings, 130.
Parallelism is not identity, 124.
Periods in the earth's history, 27.
Persistence of force. Fundamental requirements of the law of, 68.
Personal identity, 12, 13, 110, 111, 142, 145.
Philosopher's stone. The, 21.
Philosopher. The small, 25.
Philosophy. Distinction between writers on, 146.
Phrenologists : elements of truth in their teaching, 96.
Physical force differs from vital force, 67.
Physiological research can no longer be neglected by students of religious science, 88.
―― Latest results of, 124.
Physiology and metaphysics, their harmony, 117.
Pierce on the spiritual origin of force, 92.
Plastids. See *Bioplasts.*
Poetic insight. Tennyson on, 137.
Protoplasm. Beale on, 2, 3, 36.
―― Huxley on, 56.
―― See also *Bioplasm.*
Protoplasmic theory. The, 52.

RANKE, his accordance with Beale, 53.
Raphael's Madonna di San Sisto, 36.
Rationalism, its decay in Germany, 105.
Registering ganglia, 79.
Religion, how it is to be defined, 127.
Religious science, 119, 146.
Resurrection of Christ, 139, 146.
Retribution. Æschylus on, 64.

SABBATH. The : has been hated by despots, 76.
―― its Divine authority, 76.
―― its social influence and value, 75.
Salvation, in what it consists, 119.
Sankey's hymns, 127.
Sarcode theory. The, 50.
Scars on the body and the soul, 82.
Schleiden and Schwann on the cell theory, 49, 70.
Science. Alleged facts in : how they are to be tested, 51.
―― defined, 119.
―― its reasoning may fairly be tested by all scholars, 42.
―― what it owes to theology, 18.
Sciences are to be mutually tested, 106.
Scientific investigations, their relation to religion, 99.
―― questions. Right of ministers to deal with, 87.
Self-appreciation, 76.
Self-direction. The power of, 124.
Sensation. Mystery of, 122.
Senses. The : and special nerves, 84.
―― their relation to the automatic nerves, 84.
―― their testimony to the existence of the soul, 85.

Senses. Nerves of the internal, 118.
Sermon on the Mount. The, 90.
Sexual selection. Theory of, 23.
Shakespeare, his delineations of human nature, 133.
Shells. Growth of, 37.
Sleep. Awaking from, 81.
Sleeping birds. Safety of, 81.
"Smartness." American admiration for, 100.
Socrates on the immortality of the soul, 35.
―― on the nature of things, 101.
―― the service he rendered to philosophy, 18.
Soul. The : a delegated power, 84.
―― a player invested with Gyges' ring, 76-78.
―― Admission of Draper concerning, 86.
―― Agassiz on certain arguments for the immortality of the, 97-99.
―― and the brain, 36, 89.
―― competive theory of the soul-atom and the soul-fluid, 143.
―― Daniel Webster on the, 89, 90.
―― Draper on, 12, 86.
―― Emerson and Carlyle on, 87, 131.
―― God's purpose in the creation of the, 37.
―― Goethe on, 57.
―― Grecian discussions concerning the, 64.
―― is more than the will, 117.
―― is not destroyed by the dissolution of the body, 36, 85.
―― its immortality. See *Immortality of the Soul.*
―― its independence of the nervous mechanism, 76-78, 85, 93-96, 124, 141.
―― its origin, 56.
―― its personal continuance after death, 36, 85, 140.
―― its relation to the body, 12, 65, 78, 81, 91, 93-96, 109-111, 144.
―― physiological arguments for its immortality, 109-111.
―― Relations between God and, 119.
―― Socrates on the, 35.
―― Teaching of materialism concerning the, 35, 36.
―― the agent of consciousness, 85.
―― Ulrici's view of, 143, 144.
Specialists, their relation to philosophy, 120.
Species. A new definition of, 29.
―― Origin of, 4.
―― Variation of, 23.
―― Variability in, 28.
Spectroscope. The : its revelations, 27.
Spencer, Herbert, his biological theories, 65, 67, 72.
―― his definition of life, 65.
―― his fatalism, 102.
―― on natural theology, 5.
―― on the inability of chemistry to produce life, 68.
Spiritual body. Biblical teaching concerning the, 148.
―― Ulrici on the, 139-149.
Spiritual life. Laws of the, 127.
Spiritual origin of force, 16, 92.
Spontaneity of vital action. Bain's theory of the, 69.
Spontaneous generation essential to the doctrine of evolution, 19.

Spontaneous generation, if proved, would not disprove design in creation, 32.
—— no instance of it known, 13, 19, 20, 21, 27.
Sterility of hybrids, 21.
Stimulus of matter. Drysdale on the, 69, 72.
St. Mark's Tower, 66.
Strauss's "Old Faith and New," 2.
Stricker's "Histology," 69.
Sun. The: materials of which it is composed, 27.
—— Sir W. Thompson on its history, 28.
Survival of the fittest. Theory of the, 29.
Sydney Smith. Anecdote of, 76.

TENNYSON on design in creation, 98.
—— on natural law, 17.
—— on the mystery of life, 44.
Tests of truth, 106.
Theism, its safety, 16.
—— Scientific, 128.
Theological students. Duties of, 120.
Theology has a physiological side, 119.
—— The debt of science to, 18.
Thompson, Sir W., on the history of the sun and earth, 28.
—— on the origin of life, 20.
—— on the origin of living tissues, 138.
Tides. The, 107.
Tissues. Living: how they are formed, 41, 54, 70.
—— or formed matter, 38, 39.
Transcendentalism in New England, its distinctive doctrine, 132.
Trilobites. The, 29.
Truth, how it is to be reached, 119.
——Tests of, 106.
Turkish reforms, 138.
Tyndall. German estimate of, 103.
—— his Belfast address, 7.
—— his complaint of the yoke of Socrates, Aristotle, and Plato, 10.
—— his effort to change the definition of matter, 7-11, 17, 103.
—— his extension of the doctrine of evolution, 7.
—— his materialism and fatalism, 102, 103.
—— his "Musings on the Matterhorn," 7, 20.

Tyndall on the connection between matter and mind, 10.
—— on the doctrine of evolution, 10.
—— on the existence of matter, 11.
—— on the failure of materialism, 68.
—— on the materialistic atheism, 9.
—— on the origin of life, 9.
—— on the properties of matter, 20.
Tyson on the cell theory, 70.

ULRICI, account of, 143.
—— on the spiritual body, 139-149.
Unconverted. The: how they are to be reached, 127.
Uniformitarian hypothesis. The, 21.
Useless peculiarities in man and animals, 23, 24, 30.

VARIABILITY in species, 28.
Variation produced by external condition, 30.
Venice. The ancient and the modern, 46.
Verity. Tests of, 73.
Vital action. Bain on the spontaneity of, 69.
Vital crystallisation, 73.
Vital force, its relations to the immaterial principle, 144.
Vitality. Huxley on, 56.
—— life, and soul. Distinction between, 66.
Vital transformation. Frey on, 70.
Volition. Power of. See *The Will*.
Voluntary and involuntary movements, 118.

WATERLOOS of philosophical discussion. The, 47, 81.
Webster. Daniel: Anecdote of, 16.
—— Death of, 89, 90.
Will. The Freedom of the, 14, 33.
—— is an efficient cause, 11.
—— its relation to the automatic arcs, 82, 83, 84.
—— its relation to the influential arcs, 80, 81, 83.
—— Power of the, 124.
—— the fountain of force, 92.
Wundt's physical axioms, 91.
Wurtz, M., on mind and material phenomena, 12.

THE END.

For EU product safety concerns, contact us at Calle de José Abascal, 56–1°,
28003 Madrid, Spain or eugpsr@cambridge.org.

www.ingramcontent.com/pod-product-compliance
Ingram Content Group UK Ltd.
Pitfield, Milton Keynes, MK11 3LW, UK
UKHW012340130625
459647UK00009B/434